BIRDS
BORNEO
SABAH, SARAWAK, BRUNEI and KALIMANTAN

G.W.H. DAVISON and CHEW YEN FOOK

POCKET PHOTO GUIDE

BLOOMSBURY
LONDON · OXFORD · NEW YORK · NEW DELHI · SYDNEY

Bloomsbury Natural History
An imprint of Bloomsbury Publishing Plc

50 Bedford Square	1385 Broadway
London	New York
WC1B 3DP	NY 10018
UK	USA

www.bloomsbury.com

BLOOMSBURY and the Diana logo are trademarks of
Bloomsbury Publishing Plc

First published by New Holland UK Ltd, 1996, 2011 as A Photographic Guide
to the Birds of Borneo, Sabah, Sarawak, Brunei and Kalimantan
This edition first published by Bloomsbury, 2016

© Bloomsbury Publishing Plc, 2016
© text Geoffrey Davison 2016

Geoffrey Davison has asserted his right under the Copyright, Designs and
Patents Act, 1988, to be identified as Author of this work.

All rights reserved. No part of this publication may be reproduced or
transmitted in any form or by any means, electronic or mechanical,
including photocopying, recording, or any information storage or retrieval
system, without prior permission in writing from the publishers.

No responsibility for loss caused to any individual or organization acting on
or refraining from action as a result of the material in this publication can be
accepted by Bloomsbury or the author.

British Library Cataloguing-in-Publication Data
A catalogue record for this book is available from the British Library.

Library of Congress Cataloguing-in-Publication data has been applied for.

ISBN: PB: 978-1-4729-3287-7
ePDF: 978-1-4729-3285-3
ePub: 978-1-4729-3286-0

2 4 6 8 10 9 7 5 3 1

Designed and typeset in UK by Susan McIntyre
Printed in China

FSC
www.fsc.org
MIX
Paper from
responsible sources
FSC® C104723

To find out more about our authors and books visit www.bloomsbury.com.
Here you will find extracts, author interviews, details of forthcoming events
and the option to sign up for our newsletters.

CONTENTS

Introduction	4
How to use this book	5
Glossary	6
Key to coloured tabs	7
Birds in the Borneo environment	8
Field equipment	11
Where to find birds	11
Species descriptions	14
Further reading	140
Index	141

INTRODUCTION

Few books specifically deal with the birds of Borneo. The first one available was Smythies's *The Birds of Borneo*, a comprehensive treatment based upon his earlier checklist, and first published in 1960. The colour plates from that book have subsequently been used to illustrate a pocket guide. More likely to attract the birdwatcher is Susan Myers's *A Field Guide to the Birds of Borneo* (New Holland) which features fully comprehensive coverage of all species including colour artworks and distribution maps (see Further Reading, page 140).

This guide offers a range of photographs of those Borneo birds most likely to be encountered by resident and visiting birdwatchers. Also included are a variety of the endemic species and subspecies found in Borneo.

Most birdwatching has been done in the northern third of Borneo. Although there has been a recent improvement in this geographical bias, the birds of the southern two-thirds of the island are still seriously under-studied. Interesting discoveries and re-discoveries are certain to await the enterprising traveller.

Just over 580 species of birds have been found in the area covered by this book. The political terms used to define this area are used repeatedly in the text, and require some explanation. Borneo is the third largest island in the world, covering nearly 740,000 sq km. It extends from about 7°N to 4°S, and about 60% of the island is north of the equator. Borneo is split between three countries. The smallest is oil-rich Brunei Darussalam, formerly known simply as Brunei, the only country located entirely within Borneo and occupying a small part of the north-west coast. Sabah to the north-east and Sarawak to the west of Brunei are two of the 13 states of Malaysia (the other 11 states lying in Peninsular Malaysia which is treated in a companion guide in this series). Together Brunei, Sabah and Sarawak occupy the northern third of Borneo.

All the rest of Borneo is known as Kalimantan, and belongs to Indonesia. Kalimantan is divided administratively into East, West, South and Central Kalimantan, each being a large and varied territory.

This book describes and illustrates 252 of the 580 species of birds found in Borneo. They are generally speaking the commoner and more conspicuous birds to be found in each of a wide range of habitats, from mangroves along the coast to forest in the mountains. They also include, however, a variety of spectacular and interesting birds which have been covered because they characterize the bird fauna of the area and give a fuller flavour of the region's biological composition.

These 252 species therefore cover nearly all of the bird families or groups that can be found here. By familiarizing yourself with the pictures, you should be able to identify a range of birds when you first encounter them. Be warned, however, that most of the groups include additional species not illustrated here. For example, there are 12 swifts of which two are shown. More specialized books will therefore be needed if you are to identify every bird, and some of these books are listed in the section on Further Reading (page 140).

When identifying birds, it would be nice to think that comparison with a photograph is enough evidence to be sure of identification. In practice, however, a great deal depends also upon a bird's behaviour, its calls or song, and views at different angles or in different lights that reveal the full range of plumage features. Pictures alone cannot illustrate all of these features. It is therefore important, if you are to advance in birdwatching, to take notes on the birds you see and to make simple sketches of their features. From these notes and sketches it should be relatively easy to proceed to identify most birds.

Sometimes a single feature is enough to clinch identification, such as the curling red horn of the Rhinoceros Hornbill. More often, a combination of features is necessary in order to confirm a bird's identity, including size, general shape, shape and length of particular parts of the bird such as the tail, wings and bill, and colour pattern.

Much birdwatching can be done without any equipment. A notebook to jot down what you see is a basic requirement, and all except the beginning birdwatcher will require a pair of binoculars. Specialist advice on binoculars, telescopes and cameras can be found in a range of books and magazines; a few general tips are given on page 12. Most valuable of all, when identifying birds, is help from other birdwatchers. The beginner is likely to learn much more in a day spent birdwatching with an experienced friend than from a long time spent with books alone.

HOW TO USE THIS BOOK

The book has been designed with clarity and ease of use in mind. The first section describes some basic biology of birds in the different habitats represented in Borneo, with emphasis on some of the habitats that are of special significance, such as heath forest and lowland lakes. Below is a glossary of technical terms used in the book, and towards the end of the introductory pages are some notes on equipment (page 11) and on places to go birdwatching in Borneo (page 11). On page 7 is a key to the symbols used on each page of the later descriptions. These symbols are a guide to the family or group of families to which each bird belongs. Each such symbol first appears on the first full page bearing descriptions of that group.

The introduction is followed by the 252 species descriptions, which generally follow the same sequence and names as *A Complete Checklist of the Birds of the World* (see Further Reading, page 140). There have been a few changes to names based on recent knowledge, and a few changes to sequence for convenient arrangement of similar species on a single page.

The species descriptions begin with the common name, scientific name, and length of the living bird from tip of bill to tip of tail. The first sentence introduces the species to the reader. The next few sentences describe the main features of the bird needed for successful identification. This is followed by a sentence mentioning the calls and other behaviour, if distinctive and useful for identification. The next sentence begins 'Found in...' and describes the habitats in which the bird occurs. The final sentence begins 'Occurs in...' and gives the bird's distribution and its status in Borneo.

At the end of the book (page 140) there is a list of suggested further reading and an index to the species described in the main part of the book.

GLOSSARY

Carpal joint The main bend of the wing, corresponding to the human wrist

Casque A helmet-like growth on the upper part of the bill, in hornbills and a few other species, usually hollow and light

Cere Bare skin between the eye and the base of the bill

Endemic Confined to a particular region, in this case to Borneo

Fulvous Bright yellowish to rufous brown

Frugivore A fruit eater

Lores The region of the face between the eye and the bill

Juvenile Young bird in its first full plumage, which often differs from that of adults

Migrant A bird that undertakes long and regular journeys between its breeding and non-breeding areas

Montane Confined to, or specialist in, mountain habitats, locally defined as higher than about 900m

Primaries The flight feathers on the outermost portion of the wing, corresponding to the human hand, and in most birds forming the hindmost tip of the folded wing when the bird is perched

Racquets Expanded paddle-shaped tips of certain feathers

Resident A bird that remains within a general area at all seasons, not undertaking long seasonal journeys

Secondaries The flight feathers of the inner part of the wing, corresponding to the human forearm

Under tail-coverts Small feathers covering the bases of the tail feathers, beneath the tail

Upper tail-coverts Small feathers covering the bases of the tail feathers, above the tail

Wing coverts Small feathers covering the bases of the primaries and secondaries on the wing

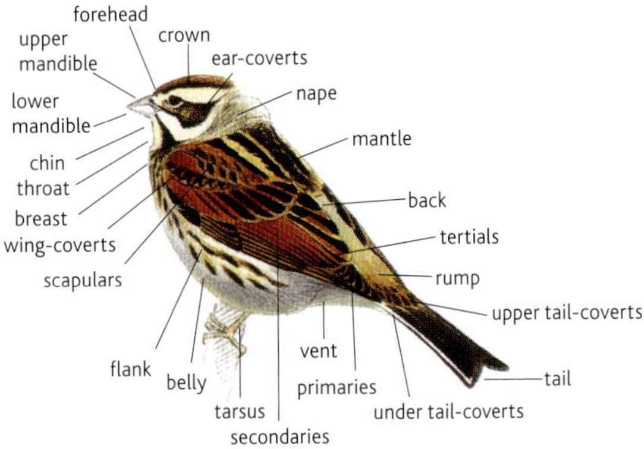

KEY TO COLOURED TABS

Cormorants, darters & frigatebirds	Herons & storks	Ducks	Raptors	Gamebirds
Rails & allies	Waders	Gulls & terns	Pigeons, doves & parrots	Cuckoos & relatives
Owls	Frogmouths, nightjars & swifts	Trogons	Kingfishers	Bee-eaters
Hornbills	Barbets & woodpeckers	Broadbills	Pittas	Swallows
Wagtails & pipits	Trillers & minivets	Bulbuls	Leafbirds & allies	Shrikes
Thrushes & relatives	Babblers & relatives	Old World warblers	Flycatchers & whistlers	Flowerpeckers
Sunbirds & spiderhunters	Sparrows, munias & white-eyes	Starlings & mynas	Drongos & orioles	Crows

BIRDS IN THE BORNEO ENVIRONMENT

Of the 630 or so species of birds recorded from Borneo, more than 300 are typical of the tropical evergreen rainforest that in prehistory covered the island and today is still one of the major habitat types. This total within the forest includes all but two of the island's 50 endemic species (the species that are totally confined to Borneo). Thus about 10% of the forest birds are endemic, and they include:

Mountain Serpent-eagle	White-fronted Falconet
Red-breasted Partridge	Crimson-headed Partridge
Bulwer's Pheasant	Bornean Peacock-pheasant
Bornean Ground-cuckoo	Dulit Frogmouth
Whitehead's Trogon	Golden-naped Barbet
Mountain Barbet	Bornean Barbet
Hose's Broadbill	Whitehead's Broadbill
Blue-banded Pitta	Black-and-crimson Pitta
Blue-headed Pitta	Bornean Bulbul
Cinereous Bulbul	Black Oriole
Bornean Bristlehead	Bornean Ground-babbler
Black-throated Wren-babbler	Mountain Wren-babbler
Chestnut-crested Yuhina	Everett's Thrush
Bornean Blue Flycatcher	Bornean Stubtail
Friendly Bush-warbler	Bornean Whistler
Yellow-rumped Flowerpecker	Whitehead's Spiderhunter
Pygmy White-eye	Mountain Blackeye

The only endemics that are not forest birds are Bold-striped Tit-babbler and Dusky Munia. Of the forest endemics, only 10 are lowland birds, while 60 per cent are montane and the rest are birds of the hill slopes that extend up into the mountains. This draws attention to the importance of the montane habitats in Borneo as they give the avifauna a special character.

The highest mountain in Borneo, and in fact in the whole region between the Himalayas and New Guinea, is Mount Kinabalu (4,101m). No species of bird is strictly confined to this mountain (the most restricted, the Friendly Bush-warbler, is also found on two neighbouring peaks), but it does support every one of the montane Bornean endemics. From Mount Kinabalu, the montane core of Borneo extends southwestwards, principally along the border region of Sabah, Brunei and Sarawak with Kalimantan. A few peaks within this main montane area have been well studied, but there has been extremely little work on any of the mountains lying within Kalimantan, from which new records of montane birds continue to trickle in.

In western Borneo, this montane massif splits into two chains of hills, one extending along the Sarawak border to end at Gunung Pueh (1,570m), and the other running south-west through Kalimantan to Mount Raja, Mount Saran and Mount Sebajan (1,377m). Between these two giant spurs lies the Kapuas river drainage, one of the biggest in Borneo. Another, smaller and much more isolated mountain range exists in extreme southern Borneo, where the highest peak is Gunung Besar (1,890m). Listing of birds from each peak has been done so patchily that few reliable conclusions can be drawn about the detailed distribution of montane birds. However, a few of the montane endemics have outlying populations on Gunung Pueh in the extreme south-east of Sarawak. This is probably an indication of a former cooler period when the distribution of montane birds in general was wider than it is today. Presumably only a few montane species have been able to hang on in such a small area of montane habitat, other species having died out in that area.

From this montane core, major rivers radiate to every coast of Borneo. Amongst the most important are the Baram and the Rajang in Sarawak, the Kinabatangan in Sabah, and the Kapuas, Barito and Mahakam in Kalimantan. Where the banks of these rivers remain forested, they can be important for birds such as Darters; a few wintering birds such as Red-necked Phalaropes occur there; and there are occasional exciting records of rare birds such as Storm's Stork and the White-shouldered Ibis. Each of the big rivers runs through a great variety of habitat types including interesting swamp forest, and the estuaries of the main rivers in Kalimantan are specially important for a variety of birds. For example, two resident species of duck were added to the Borneo list as recently as 1987, owing to studies at the mouth of the Mahakam.

Three Borneo habitats in particular deserve special mention. One is the montane forest. With increasing altitude the forest is not only cooler but damper, more stunted, mossy and shrouded in mist. The relative isolation of the Bornean mountains from those of Sumatra, Java and the Malay Peninsula has certainly been a factor in the evolution of the montane endemic birds described above. Amongst these endemics, species to look out for include the Mountain Serpent-eagle, the two spectacular montane partridges, the Mountain Blackeye and the Bornean Whistler. Slightly more widespread are montane birds that are shared with Sumatra, such as the Sunda Laugh ingthrush and the Bornean Treepie. More widespread again are those shared also with the Malay Peninsula, such as the Black-and-crimson Oriole and Greater Green Leafbird. Altogether there are approximately 62 montane specialists, fewer than in either Sumatra (85) or the Malay Peninsula (75).

A habitat of greater significance in Borneo than almost anywhere else in South-east Asia is peat-swamp forest, together with a range of other swampy and heath-like forests that grow on poor soils. This range of forest types is relatively poor in birds, but one or two species do well there. Amongst them are the Hook-billed Bulbul, the Bornean Ground-babbler, and the Grey-breasted Babbler. In tall peat-swamp forest, but also extending into lowland dipterocarp forest, is another spectacular endemic, the Bornean Bristlehead. Small flocks of this dark, crimson-fronted oddity pass noisily through the canopy and are gone. It is very difficult to photograph.

Further interest is given to the Bornean avifauna by a variety of lakes. In Sarawak, Loagan Bunut is a shallow lake now a national park. Elsewhere in Sabah and Sarawak there are only a few oxbow lakes of little significance to birds. In Kalimantan, however, there are three major lake clusters. The Kapuas lakes are more than 300km up the Kapuas river, yet they are less than 150km from the Sarawak coast and may mark a low point where the Kapuas and Lupar rivers have changed course in the geological past. Three big lakes are found 120km up the Mahakam river in east Kalimantan. Lastly, Bangkau lake in the extreme south is not associated with any major river, but was the site of collection of various rare birds that have seldom or never been seen in Borneo since. Examples are the Glossy Ibis, two cormorants, and the Comb-crested Jacana.

Man has had a close association with birdlife in Borneo since the distant past. Hunting is perhaps as intensive here as anywhere in South-east Asia. Although the number of people directly dependent on hunting for subsistence might be on the decline, hunting is still an extremely widespread part of the culture. Formerly a range of ingenious and highly effective traps was used; some of these trap-making skills are dying out, but the toll on birds and other wildlife has greatly increased through the use of firearms. Big riverside birds such as Oriental Darter, White-shouldered Ibis and Storm's Stork have been heavily affected. Some of the forest game birds are also at risk.

More recently, forest clearance for land development, as a tool to raise socio-economic standards, has been one of the factors involved in the decline of wildlife. The eastern part of Borneo has furthermore become prone to forest fires, due to a combination of logging, which has dried out the forest vegetation along access roads, encroachment by settlers, and the vagaries of dry El Niño years. There have been some studies on the effects of logging, and a little research into the effects of fires, but the results are difficult to interpret and it is hardly known whether they are site-specific effects or generally applicable.

Man's long association with birds in Borneo has been reflected in the results of archaeological digs at Niah Caves and other sites. Bird bone has been one component of the finds, though a relatively minor one that has not been studied in detail. However, the caves at which such digs have been made still show living traditions in the form of collecting birds' nests, for soup. The nests of the Edible-nest Swiftlet, and to a small extent the Black-nest Swiftlet are collected in their millions by men skilled in climbing the precarious bamboo and wooden poles erected inside limestone caves to reach the most inaccessible of rooftop nests. If the idea of eating a bird's nest that is made of saliva is stomach-churning to the unfamiliar palate, it is nothing in comparison with the activity of climbing these rickety poles, from which fatal falls are certainly not rare.

The famous caves of Gunung Mulu in Sarawak are not amongst those from which swiftlets' nests are collected. The biggest of cave passages there is more famous for its bats, but it too has its ornithological spectacle. Each evening as the bats emerge, unlucky members of the flock are picked off by magnificent fast-flying Bat Hawks. While the Bat Hawks that pursue their prey in level flight are seen by almost every visitor, only a few people notice the Peregrine Falcons which are occasionally present and which snatch their prey while diving vertically through the swirling flock of bats.

A field in which much has still to be learned is that of birds at sea. Recently the South-east Asian nations have declared extensive and exclusive economic zones, in line with the needs of oil and gas exploration and deep sea fisheries. The network of oil and gas rigs has so far yielded only a little information about migrant landbirds and seabirds; much more could still be done. On the other hand, some rocky islands have come under increasing pressure. An example is Layang-layang in the area of the Spratleys north of Borneo. Tourists can now arrive by air and find chalet-style accommodation there, but the nesting terns and boobies have definitely suffered as a result.

The past and future of birds in Borneo can therefore be seen as a very mixed package. The origin and history of the fauna is complex. A very wide range of habitats is represented, some of which are quite unusual within the region. Tremendous pressures have been exerted upon birds through hunting, nest collection and habitat change. Some research has been and is being done, but many gaps in knowledge remain, especially in southern Borneo. On the other hand, fine reserves and national parks exist, in which the future of most birds may be secure for some time to come.

FIELD EQUIPMENT

A notebook and pen should be considered basic and indispensable. Pencil smudges; ballpoint pens are good for permanency but do not work well on damp paper. Rain and humidity are problems for binoculars; you can opt to buy an expensive brand, with waterproofing, or a cheaper brand which can be replaced if spoiled. Reasonable quality binoculars of a specification around 8×30 or 10×40 are the most suitable.

Good equipment is bulky as well as expensive: specialist items such as cameras and tape recorders are not something beginners should consider until they are firmly committed to birdwatching. Most pictures in this book were taken with a 500mm or 700mm lens, using flash, at distances of up to 25m.

Wear dull colours and lightweight clothing. You can either ignore rain, or use a poncho with a hood. Leeches will be encountered in most lowland forested areas, and can be ignored by the stout-hearted or picked off before they bite. Insecticide sprays on footwear keep off most leeches but are not good for the environment or human skin. Ticks can leave their jaws embedded, and carry disease (leeches do not); use fingernails (judiciously) or a canine shampoo to remove them.

WHERE TO FIND BIRDS

Montane forest
Kinabalu Park, Sabah. The premier birding area in South-east Asia, with easy access by car or minibus and a range of accommodation and restaurants. Extensive montane forest to beyond treeline, magnificent plantlife, over 300 birds either on the mountain or in lowland Poring, including many endemics.

Lowland forest
Gunung Mulu National Park, Sarawak. A fine experience. Access by air from Miri, Marudi and Brunei, and a range of accommodation from simple to luxury in and outside the Park. Extensive lowland forest, limestone and riverine habitats; montane areas less accessible. Over 250 birds, many endemics.

Bako National Park. Access via Kuching, Sarawak. A moderate variety of birds but good for its easy access by bus/boat, and for the chance of seeing proboscis monkeys.

Niah Caves National Park. Access via Miri, Sarawak. Mainly for swiftlets, nest harvesting and caves; also forest birds in a now rather restricted forest area.

Danum Valley Conservation Area, Sabah. Access from Lahad Datu or Tawau. A fine range of lowland species in forest.

Sepilok. Access via Sandakan, Sabah. Known for its rehabilitant orang-utans, also a range of lowland birds.

Kutai National Park, east Kalimantan, access from Samarinda. Over 230 birds in somewhat disturbed lowland forest.

Tanjung Puting National Park, south Kalimantan. Access via Palangkaraya. Impoverished heath and peat-swamp forest; over 200 birds and orang-utan watching.

Coastal and marine habitats

Kota Belud Bird Sanctuary, Sabah. Access by road from Kota Kinabalu. Rice and dune slacks along coast attracting herons and other birds.

Pulau Tiga, Sabah. Access from Kota Kinabalu. Forested island with some forest species, waders, occasional passing seabirds, and notably megapodes nesting in sand.

Layang-layang islands. Access by air from Labuan. Coral islands still important for persecuted terns and boobies, other seabirds. Simple chalet accommodation.

Besides these sites, many others are protected (for example Samunsam, Lanjak Entimau, Batang Ai, Loagan Bunut in Sarawak; Crocker Range, Turtle Islands, Tabin in Sabah; Gunung Palung in Kalimantan), all of which are of high conservation value with interesting wildlife.

Acknowledgements

All the photographs in this book were taken by Chew Yen Fook with the exception of the following: David Bakewell (83a, 107b); Geoffrey Davison (30a, 61b, 61c, 117a, 117b); James Eaton (87a); Leif Gabrielsen (83b); Tan Siah Hin (21a, 24b, 36b); Taej Mundkur (15a, 19a, 37b, 44a, 45a); Nature Photographers: Robin Bush (23b), M.P. Harris (46b), Paul Sterry (43b); K.W. Scriven (31b, 74a); Ong Kiem Sian (46a); Morten Strange (105b, 136b); World Wildlife Fund Malaysia: E.L. Bennett (82b), Oh Soon Hock (35a), Lee Kup Jip (61a), Mikaail Kavanagh (29a), Dionysius Sharma (55 inset), Slim Sreedharan (42b), Edward Wong (45b). Photographs are designated a–d, reading from left to right and top to bottom.

The photographer wishes to give special thanks to Datuk Lamri Ali, Director, Sabah Parks, and Mr Francis Liew, the Deputy Director, for their generous accommodation during photography at Kinabalu Park; and to Mr Alim Bium, the Park's Technical Assistant, for invaluable advice. We are greatly indebted to Malaysia Airlines, in particular Mrs Siew Yong Gnanalingam, Corporate Affairs Manager, for generous air travel sponsorship for travel for Chew Yen Fook and Ken Scriven. We are extremely grateful to Ken Scriven for his invaluable advice, his wonderful companionship in the field, and for the use of his sound recordings, which brought in many birds that would other-wise have stayed out of camera range; and to Dr David Wells. Chew Yen Fook gives many thanks to Mr Liau Beng Cye, Director of Armourshield Marketing Pte Ltd, for his company's support, and last but not least, to his wife, Siang, for her encouragement in helping to strive for better pictures.

GREAT CORMORANT *Phalacrocorax carbo* 80cm

By far the biggest of the three local cormorants, this bird is found throughout the world. Adult is a big, heavy, black bird with a small head and longish, hooked bill, contrasting pale cheeks and throat; the upperparts browner with black scaling on wings; when breeding, has white patch on outer thigh. Juveniles are browner, with more extensive dirty white on neck and breast. Seen singly or in small groups, swimming low in the water or standing on mudflats or on bare trees, often with spread wings drying. Is found occasionally in coastal waters and on offshore islands. Occurs throughout the world; here both a resident and migrant.

ORIENTAL DARTER *Anhinga melanogaster* 90cm

Closely related to cormorants, these extraordinary birds are becoming scarce everywhere since they make easy targets for hunters in boats. Big but slender, dark, with long thin neck and dagger bill; big floppy wings often spread to dry, exposing pale markings on wing-coverts and streaked white plumes on back; tail long, fan-shaped. Distinguished from cormorants by size, tail shape, tiny head, habitat and behaviour. Perches in low dead trees, usually singly, and low vegetation along rivers; swims largely submerged with head and neck showing. Found along forested rivers from mouth to far upstream. Occurs from India to Sulawesi; resident.

CHRISTMAS ISLAND FRIGATEBIRD *Fregata andrewsi* 94cm

The first step in distinguishing the three different frigatebirds is to separate out males (with red throat patch), females (black head and white breast) and juveniles (pale buffish head). Adult male Christmas Island Frigatebird has dark breast and white abdomen. Female has black chin, white band on axillaries and white underparts extending in convex shape down to abdomen. Juvenile like female but pale head; broad, nearly complete dark breast-band, white underparts with dark scales; inner secondaries olive-brown. Typically flies in small loose flocks, showing long wings and long forked tail. Found flying offshore and on rocky islands. Occurs through eastern Indian Ocean to Indonesia, Hong Kong; occasional visitor.

GREY HERON *Ardea cinerea* 100cm

Odd individuals are found along muddy coasts, but breeding colonies are few and far between. Tall heron, bulkier than Purple Heron, grey above with black flight feathers and head patch, greyish-white below with dark streaks down central neck and breast; yellowish bill and legs. Feeds typically on mud and at water's edge, waiting for and stalking fish, frogs, crabs. Often seen perched in trees or in slow flight. Forms nesting colonies in tall trees in and behind mangroves. Locally abundant in mangrove forest and adjacent coastal mudflats, seldom inland. Occurs almost throughout the Old World temperate and tropical zones; resident.

PURPLE HERON *Ardea purpurea* 95cm

Slim and elegant when at rest, with angular appearance and snake-like neck, this heron's plumage can nevertheless be considerably bulked out on occasion. Slender, with head, neck and breast rufous-cream; black crown and stripes on face and neck. Back, wings and belly dark purplish-grey with some long rufous plumes. Typically solitary, quieter and less sociable than Grey Heron; perches on low trees and bushes, or seen motionless in shallows waiting for or stalking fish and other small animals. Scarce, in swamps and behind coastal mangroves, but feeding typically in fresh water. Occurs in most of Old World; resident and migrant.

GREAT EGRET *Ardea alba* 90cm

This is the largest of the egrets, most obviously so when in flight. Pure white, with strong bill. Non-breeding: bill is yellow, sometimes with dusky tip, face greenish-yellow, legs and toes black. Breeding: bill is black, face bluish-green, thighs reddish or greenish contrasting with rest of black legs. Waits and stalks fish and other animals in shallow water, often solitary or mixed with loose groups of other egrets. Found mainly on mangrove coasts, sometimes in rice fields and swamps. Occurs almost throughout the world; here a former scarce resident and now still a moderately common migrant.

INTERMEDIATE EGRET *Egretta intermedia* 68cm

Less abundant than the Great and Little Egrets, this migrant is intermediate in size. All white, with plumes on breast and back (but not the nape) in breeding plumage, sometimes adopted before migrants depart. Short bill and bare facial skin always yellow; legs and toes all black. Bill length and softer curve to neck distinguish from Great Egret, yellow bill colour from Little Egret. Waits for and stalks prey on mud, in grass and shallow water. Found in mangroves, mudflats, rice fields and swamps, mainly coastal. Occurs from Africa through central and southern Asia to Japan and Australia; migrant and winter visitor.

LITTLE EGRET *Egretta garzetta* 65cm

Though the commonest local egret, it seldom occurs here in such big flocks as the Cattle Egret. Pure white, with slender, sharp-pointed bill. Non-breeding: bill virtually black, and legs black with yellow (or rarely, also black) toes. Breeding: bill all black, face bluish-green; two long plumes on nape and filigree plumes on back and rump. Actively skitters while hunting fish and other small animals in shallow water. Found from coasts to inland freshwater of all types, but seldom seen in large aggregations. Occurs throughout the Old World; here it is a common non-breeding migrant.

EASTERN CATTLE EGRET *Bubulcus coromandus* 50cm

Appearing slow and sometimes clumsy in flight, most egrets can cover long distances. This species is one of the world's most successful colonists. Small egret with thick neck, pure white in non-breeding plumage with strong yellow bill and dark legs. Before breeding, head, neck, back and breast become suffused with buff, bill yellow, legs red. Often seen with cattle or buffaloes, feeding mainly on insects disturbed by them. Found especially near marshes, pools and rice fields, and in grassland. Occurs now almost throughout the world, here a nonbreeding migrant.

CHINESE POND-HERON *Ardeola bacchus* 45cm

Standing quietly at the edge of vegetation on the mud, this bird may be overlooked until it bursts into white upon take-off. Breeding plumage plain deep chestnut grading to black, with white wings, tail and abdomen. Non-breeding plumage, as typically seen in the region, light brown above with head, neck and breast strongly streaked grey-brown and white; wings, tail and abdomen white. Seen singly, at water's edge in thick vegetation, hunting small fish and insects. Found in lowlands in freshwater swamps, rice fields, lakes, sometimes along coast. Occurs throughout east and Southeast Asia; migrant.

JAVAN POND-HERON *Ardeola speciosa* 45cm

During the breeding season, this species has a marvellous golden tone to the head and neck plumes. In flight mainly white including all-white wings. Head and neck is brownish-gold, breast cinnamon, and back slaty black. Outside breeding season, light brown, streaked with dark brown and white; wings, tail and abdomen white; probably indistinguishable from the Chinese Pond-heron. Singly or in loose groups, standing on wet mud or on waterside vegetation. Found in inland and coastal freshwater bodies. Occurs from Thailand and Indochina to Sulawesi and the Lesser Sunda islands; here it is resident in south Kalimantan, occasionally non-breeding elsewhere.

LITTLE HERON *Butorides striata* 45cm

Most common in coastal regions, Little Herons are common inland as well, and the local breeding population is supplemented by an influx of migrants during the non-breeding season. Small, heavily plumaged heron, dark blue-grey or green-grey with nearly black crown; pale face markings, streaks on breast and narrow buff edges to wing feathers. Juveniles brown, stockier, streakier and less mottled than juvenile Cinnamon Bittern. Typically solitary, a stand-and-wait hunter, giving a single loud *keyaw* in flight. Found on streams, ponds, marshes, mangroves and seashore. Occurs in suitable habitat throughout the Old World tropics and subtropics; resident and migrant.

BLACK-CROWNED NIGHT-HERON *Nycticorax nycticorax* 60cm

This small, stocky heron makes impressive dusk flights to its freshwater feeding sites, from its otherwise mainly coastal habitat. Three-coloured adult has black crown and back, pearly-grey wings and white underparts. Juvenile dark brown, streaked buff below, more spotty than young Rufous Night-heron and distinguished from young Cinnamon Bittern and Little Heron by bold buff spots on back and wings, and by robust bill and stocky shape. Forms colonies, sometimes huge, in mangroves, but seldom feeds on coast, often a short way inland. Found in mangroves, mudflats, river edge and swamps. Occurs almost throughout the world; resident.

RUFOUS NIGHT-HERON *Nycticorax caledonicus* 60cm

Though long known as a non-breeding visitor, this species has now colonized Sabah and perhaps Brunei, through shifting or expansion of Philippines colonies. Adult a smooth, curiously plumbeous-brown above, rufous-buff shading to white below, with dark cap and a few white head plumes; bill and legs greenish-yellow. Juvenile is mottled brown, streaked and spotted with white, more streaky than the young Black-crowned Night-heron. Call a harsh croak, like previous species. Found in small colonies in mangrove, nipah palms in estuaries, casuarina trees on coast. Occurs from Java and Borneo to Australia and west Pacific; scarce resident in north-east Borneo, supplemented by winter visitors.

YELLOW BITTERN *Ixobrychus sinensis* 37cm

Seldom seen for more than a moment, this bird occasionally bursts up from the reeds before dropping back into the swamp. Buff plumage with creamy buff wing coverts and black flight feathers, giving bicolored appearance in flight, and black cap. Juvenile has same wing pattern, body heavily streaked brown above and below. Skulking, seen singly in dense swamp vegetation, usually alone and seeming rarer than it really is; feeding on small fish, frogs, invertebrates. Found in freshwater swamp, lotus, wet grassland, rice fields and old mining pools. Occurs throughout east Asia to New Guinea; resident.

CINNAMON BITTERN *Ixobrychus cinnamomeus* 37cm

Perhaps second commonest of the herons after the Little Heron, it is sometimes seen in flight across roads in swampy country. It is small, slim with a rich chestnut or cinnamon plumage, creamy beneath with dark streaks down centre neck and breast; yellowish bill and legs. In flight wings appear all dark, not strongly bicoloured. Juveniles darker brown, slim, with heavily streaked breast and mottled to spotted wings. A strong flier but skulking on ground. Found singly but abundant in a wide variety of mainly freshwater wetlands, feeding on small fish and frogs. Occurs throughout the east and south-east of Asia up to 2,000m; resident.

STORM'S STORK *Ciconia stormi* 85cm

A small stork with glossy body and clown-like red and yellow facial colouring, this bird has often been confused with related storks in the region. The adult has red bill and legs, bare yellow face, and white plumage on upper neck and abdomen; the remainder of head, neck, wings and body are dark brown. Seen perched singly in riverside trees, occasionally descending to the swampy forest floor in search of frogs, or wheeling high overhead. Found in lowland forest, tree-lined swamp and riverside habitats. Occurs in South-east Asia from southernmost Thailand to Sumatra, Borneo; now rare resident.

LESSER ADJUTANT *Leptoptilos javanicus* 110cm

This ugly bird, with its 'permed hairstyle that went wrong', is a scarce resident and a victim of hunting. Dark, slaty-grey wings, white underparts and upper back; massive horny grey bill and ugly, almost bare, reddish head; loose pinkish skin on neck and upper breast. Juvenile has browner tone to wings, dirtier white underparts. Seen singly or in small groups, foraging for fish, frogs, other small vertebrates and invertebrates in shallow water or on mud; occasionally soaring very high in sky. Found sparsely in coastal districts, on the shore and in rice fields near the coast. Occurs from India and south China through South-east Asia; sparse, ephemeral resident.

LESSER WHISTLING-DUCK *Dendrocygna javanica* 40cm

One of the few ducks that breeds in the region, this is a tree-hole nester whose downy young must fling themselves earthwards in their first few days of life. A rufous-brown, rather long-necked duck, with pale buff cheeks, dark brown cap, and usually some pale streaking on flanks. Occurs in small parties, diving for some vegetable and insect food, and taking floating weed from surface. Pairs rush across water, erect with one wing raised, in display. Found on old mining pools, freshwater swamps, occasionally forested rivers and behind mangroves. Occurs in South-east Asia from India to Borneo; here resident.

NORTHERN PINTAIL *Anas acuta* 55cm

Though each is individually rare, 12 species of duck occur in Borneo, more than in any nearby landmass. This species is rather long-necked, with grey bill and pointed tail. Male has brown head and neck with white line on each side of neck, white below, grey above, with pale streaks on inner secondaries. Female has slender neck; mottled brown, white abdomen. Intermediate plumage (shown) in post-breeding male; both sexes have brown speculum with warm buff bar in front. Found in fresh water in coastal districts, or on coastal mudflats. Occurs throughout northern temperate zone, migrating south in winter; an occasional non-breeding visitor to northern Borneo.

OSPREY *Pandion haliaetus* 54cm

The Osprey is found nearly worldwide but is always a special sight, perched in high, bare trees, or when flying over coast or river. Back and wings very dark brown, head and underparts white with dark markings or band across breast, and wide dark band through eye; dark tail faintly banded. Fish-eating, plucking live fish from the water and carrying them to exposed perch to eat. Usually solitary, typically seen in flight or perched on fishing stakes or bare trees. Found mainly in coastal districts and on larger rivers and lakes. Occurs worldwide, at low altitudes; migrant.

BLACK-SHOULDERED KITE *Elanus caeruleus* 32cm

Almost gull-like, this elegant and streamlined raptor is often seen from roadsides in rural areas. Adult pearly-grey, almost white on head and breast, darker on wings and long square-ended tail; has black shoulder patch. Juvenile duller, the grey back, wings and breast smudged with light brown. Seen hovering, especially in the evening, in search of rodents, or perched on exposed wires or dead trees. Found in open country, young oil-palm and edges of older oil-palm plantations, grassland and forest edge next to cultivation. Occurs from India through South-east Asia to New Guinea; resident.

BRAHMINY KITE *Haliastur indus* 45cm

This mainly coastal scavenger is one of the more common and spectacular raptors, often soaring in thermals in large numbers. Adult reddish-chestnut, with white head and upper breast. Juvenile dull brown with pale patch under wing at base of primaries. In flight, long but broad angled wings and long rounded tail distinguish it from larger White-bellied Sea-eagle and smaller, slimmer harriers. Call a thin, cat-like mew. Feeds on mudflats and open ground, especially on carrion and small prey. Found typically near coasts, especially mangroves, occasionally inland and on big rivers. Occurs from South-east Asia to Australia; common resident, especially on west coast.

WHITE-BELLIED SEA-EAGLE *Haliaeetus leucogaster* 70cm

The biggest raptor in the area (other than the odd very rare vagrant vulture), this bird's impressive wingspan alone is a good guide to identification. Adult has pure white head and underparts, pearl-grey to brownish-grey upperparts, contrasting blackish primaries and secondaries. Juvenile dark brown, with paler head, throat and base of tail, dark primaries (pale at the base) and secondaries. Confusing intermediate stages of increasing paleness, identification aided by size and sh... (Immature shown above.) Loud serial screaming. Catches f... water surface, sometimes feeds on refuse. Found mainly ...ooded rocky shores and mangroves, sometimes inland. ...dia to Australia; resident.

25

CRESTED SERPENT-EAGLE *Spilornis cheela* 54cm

As often heard as seen, this is possibly the most common of forest raptors. A robust, heavy-headed, chocolate-brown eagle. Adult has very short, square black and white crest, white spots on breast; in flight broad, rounded wings and tail with heavy black and white bars bolder than other raptors. Juvenile has pale head and underparts, becoming progressively darker with age. Yellow legs and cere. Call a piercing *a-cheee chee che!* especially in display flight. Distinguished from Mountain Serpent-eagle *S. kinabaluensis* by call and lack of black throat. Found in lowland and hill forest, soaring over forest and adjacent open country. Occurs from India and south China to Borneo and Java; resident.

EASTERN MARSH-HARRIER *Circus spilonotus* 45cm

A bird with a very wide distribution, this harrier reaches the northern parts of Borneo in winter, most of the population being young birds. Immature is dark brown, nearly black, with buff head markings and leading edge to wings; upper tail-coverts mottled dark and light. Adult male mottled dark brown, head and breast paler; in flight, secondaries grey and primaries black. Flying low over grassland and rice fields. Found in open country in lowlands and sometimes up into hill rice areas. Occurs throughout the Old World, eastern populations sometimes being treated as separate species; migrant.

CHANGEABLE HAWK-EAGLE *Nisaetus limnaeetus* 65cm

Variable birds of prey, some appearing nearly white and others nearly black. Large raptor with very broad wings, long round-ended tail, and hardly any crest. Dark phase (shown here) blackish-brown, yellow legs but grey cere, with pale flight feathers on underside of wing and grey under base of tail; thus distinguished from big-winged Black Eagle *Ictinaetus malayensis*. Light phase is dark brown above, nearly white below with several narrow bands on tail and wings. (Oriental Honey-buzzard *Pernis ptilorhynchus* is smaller, with narrower wings; tail has two dark bands at base.) Found in plantations, secondary growth, logged forest. Occurs from India to the Philippines; resident.

BLYTH'S HAWK-EAGLE *Nisaetus alboniger* 52cm

Nesting in the crotches of emergent forest trees, this spectacular crested eagle is characteristic of hills and mountains. Adult black and white, brownish tone to wings, with strongly barred underparts, long crest and broad white band across dark tail. Juvenile dark brown above with pale scales, pale head and underparts, long crest, several narrow dark bands across buff tail. Call a high quavering scream similar to Changeable Hawk-eagle. Catches arboreal mammals, birds, reportedly even bats; spends long periods quietly perched. Found in hill and montane forest where frequently seen, seldom in lowlands. Occurs only in the Malay Peninsula, Sumatra, Borneo; resident.

BLACK PARTRIDGE *Melanoperdix nigra* 24cm

This is the mystery partridge of the South-east Asian rainforest, commonly seen by a few observers at selected localities, otherwise apparently extremely rare or elusive. Male entirely black, with dark robust bill and feet. Female dark rufous-brown with pale throat and bold black bars across wings. Very secretive, perhaps partly nocturnal, giving a creaking call and reportedly a double whistle. Diet and behaviour hardly known, observations needed. Found in lowland forest over level ground, perhaps hills up to 600m, especially with stemless palms. Occurs in the Malay Peninsula, Sumatra, Borneo; resident.

CRESTED PARTRIDGE *Rollulus rouloul* 25cm

Male (above); female (right)

By far the prettiest partridge. Devoted pairs have a strong and complex family life, and congregate into flocks with their young. Male very dark, glossy blue with maroon crest, red legs and skin round eye, and bright red patch on bill; head outline distinctive. Female dull grassy green with grey head, rufous wings, reddish legs and eye-skin. Forages for insects and fruits in leaf litter, flocks working gradually through forest; call a thin, glissading whistle. Found in lowland and hill forest up to 1,500m. Occurs in the Malay Peninsula, Sumatra, Borneo; resident.

CRESTLESS FIREBACK *Lophura erythrophthalma* 45–50cm

Male (above); female (right)

The members of this species found in Borneo are more colourful than those in Sumatra and the Malay Peninsula. Male dark blue-black with plain bright buff tail, white breast streaks, lacy white markings on back and wings, and (usually concealed) iridescent orange back; facial skin scarlet. Female dark, dull blue-black including tail, greyer head, dull red facial skin. Single birds, pairs or small parties move through undergrowth, gurgling and clucking to stay in contact; males whirr wings as display. Found in logged and unlogged lowland and hill forest up to 800m, especially with stemless palm undergrowth. Occurs in the Malay Peninsula, Sumatra and Borneo; resident.

CRESTED FIREBACK *Lophura ignita* 65cm

This heavily built pheasant is hunted everywhere, but is still considered widespread. Male deep shiny blue-black, arched buff central tail feathers, rusty red abdomen, and brilliant blue facial skin. Iridescent coppery back usually hidden. Female bright chestnut-brown, scaled whitish below, becoming white on abdomen; dull blue facial skin. Both sexes have whitish flesh-coloured legs and funny bobble crest. Forages for fruits and insects in leaf litter, in noisy, clucking flocks and sometimes singly. Found in lowland forest, especially near rivers and streams. Occurs in the Malay Peninsula, Sumatra and Borneo; resident.

Male (above); female (below)

BULWER'S PHEASANT *Lophura bulweri* 50–75cm

This seldom-seen, hill forest bird has a spectacular display in which its distended wattles form a blue ribbon against the white background of its grossly spread tail. Male appears black with pure white tail and blue head; closer view shows maroon breast and rump, shining blue fringes to feathers of back and wings; black-tipped blue wattles above and below eye, with red ring round red eye, and reddish legs. Female plain brown with blue facial skin and reddish legs. Singly or in small groups, foraging on ground; call a harsh, loud *bek-kia*! Found in lowland and more often hill forest to 1,000m. Occurs only in Borneo; endemic resident.

GREAT ARGUS
Argusianus argus 75–190cm

Female (above); male (right)

One of the most famous South-east Asian birds, spectacular, always heard and seldom seen. Both sexes dark brown, markings a fine mosaic of buff, brown and black; chestnut breast, dark tail, and bare blue skin on face and neck. Male gigantic because of elongated wing and tail feathers, complex pattern with eye-spots on wings revealed in display. Small, inconspicuous crest. Males each call from cleaned display space, very loud two-note hoots, *ki-wow*! Males and females give a similar-toned series of about 30 single hoots, last few notes rising up scale. Found in lowland and hill forest to 900m. Occurs in the Malay Peninsula, Sumatra and Borneo; resident.

SLATY-BREASTED RAIL *Gallirallus striatus* 25cm

Rails are shy and secretive, keeping to dense swampy vegetation, but sometimes emerging at dawn and dusk. Adult dark grey, browner on wings, with narrow white barring all over wings, back, tail and flanks; chestnut crown, grey legs, and dark bill with red base. Juvenile similar, but browner with less obvious white barring and paler abdomen. Call a jarring double croak. Typically solitary, walking slowly on soft mud and between the stems of swamp plants or grasses; may be alarmed into flight. Found in marshes, overgrown edges of mining pools, mangroves, rice fields, thickly overgrown ditches. Occurs throughout east and South-east Asia to Sulawesi; resident.

RED-LEGGED CRAKE *Rallina fasciata* 25cm

Still more secretive than Slaty-breasted Rail, this is one of three similar-looking rails. Head and breast chestnut, back and wings brown with some white bars across flight feathers visible when bird is walking; bold wide black and white bars on flanks. Bright red legs. The similar Ruddy-breasted Crake *Porzana fusca* has no white on wing and narrow bars on flanks; the rare Band-bellied Crake *Porzana paykullii* has little white on wing and narrow black and white bars on flanks. Call of Red-legged Crake a single croak. Found in freshwater swamps and rice fields, occasionally on migration in any habitat, even forested mountains. Occurs from India throughout South-east Asia to northern Australia; resident, migrant and winter visitor.

WHITE-BREASTED WATERHEN *Amaurornis phoenicurus* 33cm

Bamboo flutes are used by hunters to attract this bird within range, as it advertises its presence with wild, eccentric gurgles and clucks for minutes on end. Adult has white face, breast and abdomen, chestnut flanks and dark grey (nearly black) crown and upperparts; greenish legs, mainly yellow bill, and pale rufous under tail-coverts. Juvenile similar but duller, paler, dirtier-looking. Seen crossing roads, flying out of swamps or grassland, often accompanied by black fluffy chicks. Found in agricultural estates, secondary growth, ditches, rice fields, swamps and mining pools. Occurs from India to the Philippines; resident and migrant.

WATERCOCK *Gallicrex cinerea* 42cm

Bigger and heavier than most of the rails and crakes, watercocks seldom stay in view for long. Here typically in non-breeding plumage: bulky, buffish-brown, with dark brown to blackish streaking above and narrow barring on breast and flanks; dull green bill and legs. Breeding plumage (male): blackish with brown scaling on wings and back, rufous beneath tail; red legs and mainly red bill and shield on forehead. Many intermediate plumages seen, seldom in full breeding plumage here. Similar range of eccentric calls to White-breasted Waterhen but deep, booming, brief. Found in marshes, rice fields, behind coastal mangroves. Occurs throughout east Asia; migrant.

COMMON MOORHEN *Gallinula chloropus* 32cm

The bobbing motion and cocked tail when swimming are characteristic of this cosmopolitan bird. Adult slate-grey to black with broken white stripe on flanks and two white patches separated by black line beneath tail; red forehead shield and bill with yellowish tip. Juvenile dull dark grey-brown, dusky whitish below with white patches beneath tail; dull greenish legs and bill. Call a single musical croak; often seen swimming, occasionally walking, often several to many in open water, swimming with raised tails. Found on mining pools, rice fields, swamps. Occurs worldwide, except the poles and Australia; migrant; may be resident in south.

BLACK-BACKED SWAMP-HEN *Porphyrio indicus* 42cm

The sight of these birds, big yet remarkably inconspicuous, is one not to be missed. Adult unmistakable bright bluish-purple, darker blue-black on the wings and head; big red bill and legs, white beneath tail. Juvenile duller with blackish bill. Call a varied harsh cackling and grunting, poorly described; seen usually in ones or twos at edge of water in lotus, reeds, grasses; feeds mainly on plant matter, some invertebrates, often holding food in foot. Found in rice fields, swamps, old mining pools. Occurs on Sumatra, Java and Sulawesi; resident in south Kalimantan. Formerly regarded as a race of Purple Swamp-hen *Porphyrio porphyrio*.

COMMON COOT *Fulica atra* 35cm

A variety of waterbirds including this species, as well as ducks, rails, crakes and lapwings, reach Borneo as occasional vagrants. Slaty black all over, adult with a white frontal shield on forehead and reddish eye; legs greenish, the feet not webbed but the toes with lobes of skin down each side. Juvenile browner, paler below. Distinguished from Common Moorhen by lack of white on under tail-coverts. Swims on open water, when head jerks back and forth, or amongst dense vegetation. Found along margins of freshwater bodies, swamps, occasionally along coast. Occurs throughout Europe, Asia and Australasia, dispersing outside breeding season; a rare vagrant to Sabah.

PHEASANT-TAILED JACANA *Hydrophasianus chirurgus* 30–55cm

Elegant, elongated, with very long toes and claws; females are brightly coloured whilst males care for nest and young. Non-breeding, tricolor appearance: brown back and short tail, white beneath and on spread wing, black wing-tips, crown, neck-stripe and breast-band; buff hind neck. Breeding: mainly black with white wing-patches, face and neck; black neck stripe and yellow hind neck; long tapered black tail. Call a high-pitched repetitive nasal note; usually seen here singly, walking delicately over mud. Found in mangroves, swamps, overgrown mining pools especially with lotus. Occurs from India to the Philippines; resident in south Kalimantan.

GREATER PAINTED-SNIPE *Rostratula benghalensis* 25cm

One reason this bird is seldom recorded in Borneo is its extremely secretive habits, and it is active to some extent by night. Adult female has rich chestnut-maroon forequarters, sharply cut off by half-collar from the white lower breast and abdomen, and by buff scapulars from the dark wings; white mark round eye. Male a washed-out version of female, the forequarters greyish and eye marking buff. Alone or in pairs, superbly camouflaged. Found in swamps, thick reeds and lotus, rice fields, water hyacinth in lowlands. Occurs from Africa to Japan and Australia; a migrant and possibly scarce resident.

BLACK-WINGED STILT *Himantopus himantopus* 38cm

Careful notes should be taken of stilts seen in Borneo, to distinguish between southern migrants with blackish nape, and northern migrants with white or grey nape. Etiolated, with long red legs, black bill, black back and wings. Underparts and most of head and neck white. In flight, solid dark wings and white head and rump, with trailing feet distinctive. Noisy piping *chip chip chip* in flight; typically seen in small groups separate from other waders, feeding in deep water by wading and swimming. Found in freshwater swamps, coastal mudflats. Occurs through most of the Old World; here a rare visitor.

PACIFIC GOLDEN PLOVER *Pluvialis fulva* 25cm

Non-breeding: speckled brown, buff and gold above, appearing brown at a distance, with buffish breast, eyebrow, face and buffish, abrupt forehead. Breeding: upperparts become brighter gold, face and underparts black, with white band from forehead over brow down sides of breast to flanks. Intermediate plumages with some black below. In flight, plain brown above (no white anywhere), dirty light buff below. Lack of black 'armpit' patch in flight separates it from the larger Grey Plover *P. squatarola*. Call a shrill *keruit*. Found commonly on short grassland, rice fields, muddy coasts. Occurs in most of Asia; migrant.

LITTLE RINGED PLOVER *Charadrius dubius* 17cm

A hunched and hesitant runner on mud or sand, this little plover has a rapid direct flight low over the ground. Small, with brown upperparts, white below; black mask extending into patch on forehead with white above and below it, white collar and black breast-band; legs and eye-ring yellow. Juvenile has black replaced by brown, overall duller, obscurely marked. In flight small, no wingbar, white edges to tail, call a long-drawn, descending whistle. Found on coasts, mudflats and sand bars, short grassland and riverbanks. Occurs from Africa through Asia to New Guinea; migrant and non-breeding visitor.

LESSER SAND PLOVER *Charadrius mongolus* 20cm

Here, Lesser Sand Plovers (above centre, upper r., lower r.) are with Curlew Sandpipers (longer curved bill, mottled breast, scalloped wing coverts). Lesser Sand is small with short bill and dark grey legs, sewing-machine feeding action; Greater Sand Plover has robust bill, green-grey legs, deliberate feeder. Non-breeding: sandy brown above and on patch at side of breast; white below and on forehead-cum-eyebrow. In flight, narrow white bar along wing, white-bordered tail, call pipip. Adopts partial breeding plumage before leaving, with broad rufous breast-band, white forehead with black above it and through eye. Found on muddy and sandy coasts, seldom inland. Occurs throughout the Old World; migrant and non-breeding visitor.

BLACK-TAILED GODWIT *Limosa limosa* 40cm

There are two godwits, both tall and straight-billed, of which this is the commoner. Non-breeding: greyish-brown above, faintly mottled; dusky whitish below and on eyebrow, with black-tipped pink-based bill, black-tipped tail, white rump and (in flight) striking white wingbar. Breeding: becomes rufous on head, neck and breast, barred on flanks and abdomen. Bar-tailed Godwit *L. lapponica* is more mottled above, with barred tail and almost no wingbar; in breeding plumage more deeply and extensively rufous. Both species occur on coasts and mudflats; Black-tailed also sometimes inland. Occurs throughout the Old World; migrant and non-breeding visitor.

WHIMBREL *Numenius phaeopus* 44cm

By far the commonest of four species of curlew occurring in Borneo. Small, dark brown, with less curved bill and strong blackish and buff stripes on crown; in flight has pale off-white rump smaller, less obvious than in the Eurasian Curlew *N. arquata*; wing plain above, pale below with heavy dark barring. Call a one-tone multiple trill not ascending in scale; common feeder out on mudflats, coming to mangroves at high tide and at night. Found on all muddy coasts, swamps behind mangroves, in mangrove trees. Occurs almost worldwide; migrant and non-breeding visitor.

COMMON REDSHANK *Tringa totanus* 28cm

One of the easier waders to identify, this species is found throughout the Old World. Non-breeding: grey-brown above and on head and neck; red legs and red bill with black tip; distinguished from Spotted Redshank *T. erythropus* by more uniform upperparts and fainter eyebrow. Breeding: darker, increasingly mottled above, heavily streaked below. In flight, broad white patch from inner primaries across entire secondaries, white rump, narrowly barred tail. Gives three-note or four-note piping call with first note most emphatic. Found on muddy coasts, occasionally swamps inland. Occurs throughout Europe, Africa, Asia to Sulawesi; a common migrant and non-breeding visitor.

COMMON GREENSHANK *Tringa nebularia* 35cm

This is the second commonest of the larger sandpipers, after Common Redshank. Non-breeding: light grey mottled above, on head and sides of neck; whitish below, with long green legs and robust, straight black bill. Breeding: more heavily mottled. In flight, plain wings, white rump extending up back, barred tail beyond which feet project moderately. Superficially like Marsh Sandpiper *T. stagnatilis* but large size, robustness, loud ringing three-note call in practice make separation easy. Often seen in big flocks, wheeling and settling. Found on coastal mudflats, rice fields, swamps. Occurs throughout the Old World and the Pacific; migrant and non-breeding visitor.

WOOD SANDPIPER *Tringa glareola* 23cm

The squared-off head and chequered wings of this freshwater wader are often sufficient to clinch its identification. Upperparts grey-brown, strikingly freckled, breast greyish, head dark with pale eyebrow; yellowish legs. In flight, square white rump contrasts with dark back, underwings pale near body, no wingbar. Typically noisy, *chiff chiff chiff* and more varied notes in flight; feeds around freshwater margins on insects and insect larvae. Found in swamps, rice fields, old mining pools, mangroves; one of the commonest freshwater waders. Occurs throughout the Old World; migrant and non-breeding visitor.

39

TEREK SANDPIPER *Xenus cinereus* 25cm

Typically found in small flocks which mix in with other species when feeding, and commonest on mudflats. Long, slightly upturned black bill with yellow base, bright yellowish-orange legs; upperparts dark grey faintly mottled, pale eyebrow, rump only a little paler than back; white below. Wing with broad white panel on secondaries is distinctive in flight; when standing, white panel is concealed, but dark carpal joint may be seen. Call a quick high-pitched *yee yee yee*. Found on coastal mudflats, typically in small groups or singly, mixed with other wader species. Occurs throughout the Old World when not breeding; migrant.

COMMON SANDPIPER *Actitis hypoleucos* 20cm

Abundant but typically solitary, this species skims low over the water on down-bowed wings. Sandy brown above and on crown, with light brown patches either side of the breast; underparts are otherwise white, eyebrow pale. In flight, shows striking white wingbar and white bars confined to sides of the tail. Exhibits teetering motion with bobbing tail while feeding along the water's edge, call a high-pitched three-note piping on taking flight. Found in dune slacks, river edge, rice fields, swamps, mangroves. Occurs throughout the Old World; a common migrant and non-breeding visitor.

RUDDY TURNSTONE *Arenaria interpres* 23cm

This is an attractive short-legged wader with complex head and wing markings, usually seen in small numbers mixed with other species. Non-breeding: mottled dark brown on head, back and wings, the breast darker; lower breast and abdomen white. Orange legs, short black bill and abrupt forehead. Breeding: complex pied head and neck pattern, increased plain rufous on back and wings. In flight a white triangle and narrow white wingbar, white back separated by dark rump from white tail tipped black. Trots around on mud flicking over weeds, pebbles, in search of food. Found on mudflats. Occurs worldwide; migrant and non-breeding visitor.

PINTAIL SNIPE *Gallinago stenura* 25cm

The commonest of the three snipe in Borneo. Usually seen in brief, direct flight showing little or no white trailing edge of secondaries, no white on underside of wing, with feet projecting well beyond tail. Common Snipe *G. gallinago* has white beneath and on trailing edge of wing and erratic flight; Swinhoe's Snipe *G. megala* has heavier flight and feet project little. On the ground, very long-billed sandy brown waders with head stripes and pale braces. Found in ditches, swamps, rice fields, edges of old mining pools in lowlands. Occurs through east and south Asia to Sulawesi; a common non-breeding migrant.

ASIAN DOWITCHER *Limnodromus semipalmatus* 34cm

This wader like a miniature godwit is an exciting find for most when it is picked out by thorough examination of mixed flocks. Non-breeding: dull grey-brown above, mottled like Bar-tailed Godwit but smaller, with very long, bulbous-tipped bill held downwards; underparts finely mottled grey. Breeding: back browner, face, breast and abdomen rufous. In flight, vague whitish wingbar and rump with fine darker markings. Usually silent; feeds by pivoting body on stiff legs, neck stiffly forwards, deliberately probing deep with bill. Found on muddy coasts and estuaries. Occurs through east Asia to north Australia; migrant and non-breeding visitor in small numbers.

RED-NECKED STINT *Calidris ruficollis* 16cm

This is the most easily recognized of three tiny waders that present identification problems in non-breeding plumage. Pale scaly grey above; whitish below, on face and eyebrow, with sides of breast pale grey. Short dark bill, dark legs. Becomes rufous in breeding season. In flight shows white wingbar and pale sides to rump. Related Temminck's Stint *C. temminckii* is browner with yellowish legs, browner breast, brighter white sides to tail. Long-toed Stint *C. subminuta* is taller, more heavily scaled above, with yellowish legs, and wingbar very narrow to absent. Found mainly on coastal mudflats, the other two species more on freshwater. Occurs from east Asia to Australia; a non-breeding migrant.

CURLEW SANDPIPER *Calidris ferruginea* 22cm

Recognized by its unusual bill shape, and by the particoloured individuals about migration time, this was one of the commoner shorebirds. Non-breeding: scaly pale grey above, whitish below and on eyebrow; dark legs and dark decurved bill. In flight, square white rump and narrow wingbar. Breeding: bright chestnut on face and underparts, back scaly brown. Many individuals are seen in attractive partial breeding plumage before departure each year. Clean, slim, thin-legged appearance and bill shape characteristic. Found on coastal mudflats, occasionally on swamps inland. Occurs through most of the Old World; a non-breeding migrant.

BLACK-HEADED GULL *Larus ridibundus* 36cm

All gulls so far seen in Borneo have been of this species. Non-breeding plumage: white, wings pale grey above with outer primaries forming a white wedge narrowly tipped black; head white with black smudge behind eye; undersides of primaries blackish. Juvenile has similar wing pattern but mottled with brown, and black line along rear edge of wing; tail with black band near tip. Some adults attain breeding plumage before departure, head chocolate-brown, bill and feet red. Found along coast, on mudflats, occasionally up larger rivers. Occurs throughout Europe and Asia to New Guinea; a non-breeding migrant to Sarawak, Brunei, Sabah, November to May.

WHISKERED TERN *Chlidonias hybrida* 27cm

One of the 'river terns' which patters over the water surface when feeding, rather than diving. It may be overlooked through confusion with similar species. Cold, dusky grey appearance; in non-breeding plumage a streaked blackish cap on hind crown, forehead white; upperparts uniform grey, underparts white, tail shallowly forked. Distinguished from White-winged Tern by lack of white collar, and ear patch blending into dark cap. When breeding, complete black cap, dusky blackish breast, white sides of face. Found along coasts, sometimes in inland swampy areas close to the coast. Occurs almost throughout the Old World; an occasional non-breeding migrant.

WHITE-WINGED TERN *Chlidonias leucopterus* 25cm

This is the only tern that can be found far inland, on rivers and inland lakes. In non-breeding plumage the streaked blackish cap is separate from black smudge on ear coverts; white collar separates cap from uniform grey wings, and rump is slightly paler than back; underparts white, tail shallowly forked. When breeding, head and body black, including undersurfaces of wings, and wing coverts pearly whitish. Found at sea, in estuaries, rivers, inland pools and lakes including flooded fields. Occurs through most of the temperate Old World to Australia; non-breeding migrant.

BLACK-NAPED TERN *Sterna sumatrana* 30cm

This is mainly an offshore tern, which can be seen either from boats or on rocks. Very white appearance, with long deeply forked tail, thin black bill and legs, and black band extending from eye round back of head. The juvenile has head-band less clean-cut, an obscure dark carpal bar, and upperparts scaled darker. In flight look for narrow wings, white appearance, tail shape, and very little grey on the outer primaries. Found on offshore rocky islets, open coast and mudflats. Occurs through the Indian and Pacific Oceans south to Australia; here a resident on islets, and a non-breeding visitor.

SOOTY TERN *Onychoprion fuscatus* 40cm

A maritime tern, occasionally turning up round all the coasts of Borneo but mainly in the south and east. Adult blackish-brown above, white below, with white forehead, white leading edge to inner wing, and white sides of tail; black line from bill to eye. Outside breeding season, the crown and lores streaked white. Distinguished from Bridled Tern *O. anaethetus* by lack of pale collar and darker back and wings. Found mainly on the open sea, occasionally coming inshore or blown in by storms. Occurs throughout tropical and subtropical seas; here non-breeding dispersant mainly during northern autumn.

LITTLE TERN *Sternula albifrons* 22cm

Individuals of this little bird have been seen almost every year, but they have never been proved to breed in Borneo. Tiny, grey above and white below, with black cap separated from bill by white forehead. Breeding: feet yellowish, bill yellow with black tip. Non-breeding: feet and bill blackish. Juvenile similar but some obscure scaling above and dark patch on carpal joint. In flight tiny, often hovering, black panel formed by outer primaries. Found on sandy beaches, estuaries, mudflats, coastal ponds. Occurs almost worldwide; migrant in small numbers to Borneo, and has bred in Brunei.

GREAT CRESTED TERN *Thalasseus bergii* 45cm

This bulky tern needs to be distinguished with care from Lesser Crested *T. bengalensis* and very rare Chinese Crested Tern *T. bernsteini*. Back and rump grey, underparts white, with black cap when breeding, blackish mark behind eye to nape when not breeding; short bushy crest. Yellow bill and black legs; Lesser Crested is slimmer, lighter, with orange bill and paler grey wings; Chinese Crested has black-tipped yellow bill and paler wings. In flight, obscure dark mark along trailing edge of primaries beneath wing. Found in coastal waters. Occurs discontinuously from the Indian Ocean to Australia; an occasional non-breeding visitor.

SPOTTED DOVE *Streptopelia chinensis* 30cm

A largely ground-feeding bird, the most abundant pigeon of open country and cultivation. Brown and mottled darker brown above; vinous pink head and underparts, with white-spotted black half- collar joined round sides and back of neck. Taking off, conspicuous whitish sides to tail and pale grey panel on carpal joint of each wing. Seen singly or in pairs, feeding on seeds and other vegetation, or perched on wires or low trees; not in large flocks. Call a three- or four-note *coo*. Found in oil-palm and rubber, scrub, secondary forest, villages and gardens. Occurs from India through South-east Asia, introduced through to New Zealand and in America; resident.

LITTLE CUCKOO-DOVE *Macropygia ruficeps* 30cm

A rapid frog-like note, endlessly repeated, draws attention to this, the commonest of all pigeons in montane forest. A slim, brown, long-tailed pigeon with rufous head, faintly or not barred above and below. Male has green iridescence hard to see on sides of neck; female has some dark scaling on breast and upper back. Call a rapid *wuck wuck wuck*, resolving at close range into *kuwuck kuwuck kuwuck*, about two notes per second, for minutes on end. Feeding on small fruits at forest edge; nests often found in dense low growth such as ferns. Found in montane forest, sometimes down to 500m or even in lowlands. Occurs throughout South-east Asia; resident.

EMERALD DOVE *Chalcophaps indica* 25cm

An iridescent forest pigeon that also survives well in other tree-covered habitats. Seen well it is finely coloured, iridescent green on back and wings, rich vinous pink on face and underparts, with two whitish bars across rump. Male overall brighter than female, with ashy crown and clear white eyebrow; red legs and bill. Call a soft *tik-cooo*, repetitive, first note inaudible at a distance. Found on ground, singly or in pairs, feeding on fruit and seeds in oil-palm and rubber, secondary growth and lowland forest up to 1,200m. Occurs from India to Australia; here resident but with long-distance movements.

ZEBRA DOVE *Geopelia striata* 20cm

The plumage of this tiny dove is inconspicuous against its sandy habitat; the bird will usually fly off only when the observer comes very close. Pale grey-brown all over, palest on head, the plumage barred and scalloped with narrow black lines. Long, tapered tail with white corners seen on take-off. Feeds on the ground, especially on sandy ground, singly or in pairs; call a high, far-carrying *ka-do-do-do-do* or *kaddle-a-do-do-do-do*, making it a popular competition cage-bird. Found on road margins, in gardens, agricultural estates, grassland and coastal scrub. Occurs from Burma to Java, recently split from Peaceful Dove *G. placida* of Australia; resident.

NICOBAR PIGEON *Caloenas nicobarica* 40cm

This most unusual pigeon is ungainly on the ground but is a swift flier. Dark iridescent green, becoming blackish on head and wings, with a white tail; the green feathers of forequarters lengthened into pointed golden-green hackles; knob at base of bill; short ugly reddish legs. Juveniles are duller, with dark tail and no hackles. A wary bird, feeding on the ground, occasionally in trees, found on small offshore islands. Occurs from Andaman to the Solomon Islands; resident, not yet recorded from mainland but presumably wanders between islands.

LITTLE GREEN-PIGEON *Treron olax* 20cm

Less common than the Pink-necked Green-pigeon, this species tends to be found in smaller flocks and to be more characteristic of the foothills. Both sexes have dark tail with broad grey tip. Male has grey head, orange breast-band, extensive maroon wings and back. Female has grey head, pale throat and green breast. Distinguished from others by small size, small dull bill. Call a soft undulating or lilting coo, rather brief. Feeds in small flocks on small ripe fruits including figs, at forest edge and within forest. Found from lowlands to 1,100m in forest, secondary growth. Occurs in the Malay Peninsula, Borneo, Sumatra, Java; resident but with long-distance movements.

49

PINK-NECKED GREEN-PIGEON *Treron vernans* 27cm

This chubby pigeon can be enormously common in coastal scrub when fruits are abundant. Distinguished from other green-pigeons by grey tail with black band and grey tip. Male has grey head passing through pink to orange lower breast; green back and wings. Female dull green without markedly pale throat, best identified by tail pattern and association with distinctive male. Call a lilting coo, varied and prolonged. Found in flocks especially near coast in mangroves, scrub, secondary forest, at forest edge. Occurs from southern Burma to Sulawesi and the Philippines; resident but with long-distance movements.

THICK-BILLED GREEN-PIGEON *Treron curvirostra* 27cm

An attractive, sociable pigeon; big flocks can sometimes be found at fruiting trees in lowland and hill-forest habitats. Bright yellow-green bill with red base, red feet, blue-green skin round eye. Both sexes have grey cap and grey tip beneath tail. Male has maroon wings and back (green in female) and chestnut under tail-coverts (streaked in female). Flocking, feeds on figs and other fruits; look for bright thick bill and eye. From mangrove coasts up to 1,100m in lower montane forest, fairly common inland. Occurs from the Himalayas to the Philippines and Java; resident.

JAMBU FRUIT-DOVE *Ptilinopus jambu* 27cm

On hilltops this is a surprisingly common night-flying migrant, but it is difficult to find by day. Male dark green above, carmine head, ivory-white below with pink centre to breast. Female duller dark green all over except for dull carmine face, pale abdomen, buff beneath tail. Both sexes have orange-yellow bill, red legs, pale ring round eye. Despite brilliance, often quiet and inconspicuous; Males sway from side to side in display, not bowing. Found in primary and secondary forest from lowlands to hills up to 1,100m. Occurs in the Malay Peninsula, Borneo, Sumatra; resident but with long-distance movements.

GREEN IMPERIAL-PIGEON *Ducula aenea* 42cm

Probably always patchily distributed, these heavy birds can be quite difficult to locate except when calling as they feed, taking fruits up to the size of nutmeg. Soft ashy grey with chestnut under tail-coverts, all-dark tail; green upperparts are not striking. Red feet and red base to grey bill. Now reduced by hunting to few sites (Endau Rompin, various islands, mangrove coasts), especially riverine forest, feeding high in canopy deeply cooing in small flocks. Found in lowland forest, mangroves, coastal areas. Occurs from India to south China and New Guinea; resident but local, disrupted by habitat fragmentation.

MOUNTAIN IMPERIAL-PIGEON *Ducula badia* 46cm

The biggest of pigeons in Borneo. Individuals or pairs can occasionally be seen far above the forest canopy flying down from the mountains. Grey with pale chin, maroon back, cream (not chestnut) under tail-coverts and dark tail with broad grey band on tip. Identifiable even in flight by size and tail-band; call a distinctive, very deep *whomp whoomp*, repetitive, the second note deeper and louder. Solitary or in small parties feeding in canopy or flying high over forest; display flight flapping steeply upwards to stall and glide down. Found in montane forest, often dispersing to lowlands or coast daily to feed. Occurs from India to Borneo, Java; resident with long-distance movements.

PIED IMPERIAL-PIGEON *Ducula bicolor* 40cm

This distinctive coastal and island pigeon has been a target for hunters but can still occur in huge flocks. Creamy to ivory-white, often sullied by food, with black flight feathers and tip of tail; bill and feet blue-grey. Juveniles are greyer. Form big breeding colonies on offshore islands, where fig trees are abundant; call a noisy clucking *hoo-hoo-hoo*. Also seen singly or in small groups especially at dawn and dusk over coast. Found on small islands, in mangrove and other coastal forests. Occurs from the south of Burma to New Guinea; resident species but it undertakes long-distance movements and interchange between islands.

BLUE-CROWNED HANGING-PARROT *Loriculus galgulus* 14cm

This tiny, acrobatic hanging parrot seeks fruit at the tips of twigs, and is said even to sleep inverted. Very small, brilliant green with scarlet breast and rump patches (puffed out when in display), an orange-yellow back and black bill. The blue crown is inconspicuous. The female is green with black bill and red rump; faint indications of orange-yellow back may be visible. Typically seen in small groups, high up in the canopy, moving from tree to tree. Found in lowland forest up to 1,100m. Occurs in the Malay Peninsula, Borneo and Sumatra; resident.

LONG-TAILED PARAKEET *Psittacula longicauda* 42cm

This is the most widespread of the few parrots found in Borneo; it nests in tree holes from which the young are sometimes taken as pets. Sage-green, bluer on back, wings and long narrow tail. Male with red bill and face, black chin; female duller with brownish bill, blackish chin and line through eye, no blue on back. Flies rocket-like over trees, giving single screeches; underside of wings yellow. Feeds in small parties. Found in oil-palm estates, lowland and hill forest, and secondary growth. Occurs from the Andamans through the Malay Peninsula, Borneo, Sumatra; resident.

CHESTNUT-WINGED CUCKOO *Clamator coromandus* 45cm

This unusual migrant cuckoo can with luck be seen in flight in rural areas, when its plain chestnut wings are distinctive. A slender, long-tailed cuckoo with black upperparts and tail, short crest bordered behind neck by a white collar; rufous throat and upper breast grading into a white abdomen; chestnut wings. It is distinguished from coucals, which also have chestnut wings, by slender build, long graduated tail, black back and pale collar. Solitary; call an unobtrusive throaty whistle *kooup*. Found in forest edge and water margins, secondary vegetation, rubber estates. Occurs from India to Sulawesi; migrant.

LARGE HAWK-CUCKOO *Hierococcyx sparverioides* 35–40cm

Two subspecies of this bird, one resident and one migrant, have distinctly different calls. Adult has dark grey head, face and chin, dark brown back, long, heavy, rounded tail with several broad cross-bars and a narrow whitish tip; breast rufous, flanks and abdomen strongly barred dark brown on white. Migrants, found in lowlands, are larger with dark streaks on the rufous breast, give three-note call *brain-fe-ver*; residents are montane, smaller with plainer breast and leave off third note; call is repetitive, ascending in scale, and maddening. Found up to 2,000m in forest, secondary growth, mangrove. Occurs from the Himalayas to Sulawesi.

BANDED BAY CUCKOO *Cacomantis sonneratii* 22cm

Most cuckoos, including this species, are difficult to spot as they tend to sit still in dense vegetation whilst calling. A small cuckoo, rufous-brown above, on crown and mark behind eye; whitish eyebrow and underparts, overall finely barred blackish. Rather long bill, fineness of barring and frequently erect posture help identification. Call a quick four-note *tea-cher-tea-cher*, the first and third notes the more emphatic. Found in inland forest and coastal dryland forest, secondary growth and broken forest fragments interspersed with scrub and grassland; mainly lowlands, sometimes hills. Occurs from India to the Philippines; resident.

ASIAN KOEL *Eudynamys scolopacea* 42cm

Male (left); female (above)

The scarcity of crows and mynas in Borneo, as suitable hosts to rear the young, may be one reason why Koels have been slow to colonize this region. This is a big, long-tailed cuckoo, the male glossy black all over with red eye and pale heavy bill, the female similar in shape but dark brown with bold spots and bars all over. Calls unmistakable, often starting before dawn: ten or more increasingly loud glissading notes, *koel*! Also a harsher, quick bubbling call, similar in tone, kwow-kwow-kwow-kwow. Found in mangroves, secondary growth, gardens, and town parks. Occurs from India to Australia; a migrant.

55

CHESTNUT-BELLIED MALKOHA *Rhopodytes sumatranus* 40cm

Of the five Borneo malkohas, this and the Black-bellied are particularly similar in appearance; they can be tracked down by their slow ticking or clucking from dense vegetation. Dark grey all over but for dark, dull chestnut abdomen and under tail-coverts, and white tips to tail feathers; greenish bill and red skin round eye. Black-bellied Malkoha *R. diardi* is smaller without chestnut abdomen. Forages in middle storey, giving spaced single ticks and soft mewing calls. Found in forest edge, secondary growth and mangroves. Occurs from south Burma to Sumatra and Borneo; resident.

RAFFLES'S MALKOHA *Rhinortha chlorophaeus* 33cm

The voice of Raffles's Malkoha is quite reminiscent of some trogons; malkohas are more active and forage through the vegetation in a characteristic way. Small. Male is bright gingery brown with barred blackish tail, each feather being tipped white; blue-grey skin round eye. Female similar but with light grey head, tail bright rufous-brown, and each feather tipped black and white. Call three to six thin, slow notes, descending in scale. Feeds in middle storey, peering under leaves for big insects. Found in lowland forest, secondary growth. Occurs from south Burma to Sumatra and Borneo; resident.

CHESTNUT-BREASTED MALKOHA
Zanclostomus curvirostris 45cm

In Borneo this, the most attractive of the malkohas, has more extensive rufous on the sides of the face than in the Peninsular Malaysian individual shown here. Iridescent dark green above, dark tail without white feather tips, and underparts entirely dark chestnut; red facial skin and heavy greenish bill. Seen singly or in pairs, creeping through dense vegetation, gradually working upwards within one tree then gliding down to next; spaced regular ticking or knocking call. Found in middle storey of lowland forest to 900m, plantations, secondary growth. Occurs from south Burma to Java; resident.

GREATER COUCAL *Centropus sinensis* 52cm

Coucals are big, floppy cuckoos with a weak and lumbering flight, and they quite often come down to the ground. This one is large and heavy-moving with bright rufous back and wings; rest of plumage black including head, underparts and long loose tail. Heavy horn-coloured bill. Juvenile similar but finely barred grey on black areas, and barred blackish on wings and back. Call a deep, far-carrying *hoo* or *boot*, from a few to many notes, delivered slowly, first descending then gradually rising in pitch. Found in scrub, grassland, secondary vegetation, edges of mangrove, riverbanks. Occurs from India to the Philippines; a common resident.

57

LESSER COUCAL *Centropus bengalensis* 37cm

Less easy to see than its bigger relative, this species can be located best by its call, often heard in rural scrub. It has the same colour pattern of black plumage with chestnut back and wings, but typically with some pale streaking on plumage, small bill, smaller and less lumbering appearance. Juvenile rufous-brown, paler below, with many pale buff streaks above and below; tail barred. Intermediates are common. Call a harsh grouse-like double note, *kok-kok*, repeatedly; and a short series of *hoop* notes delivered quickly and descending in pitch. Found in scrub, grassland, secondary vegetation, swamps. Occurs from India to east Indonesia; resident.

REDDISH SCOPS-OWL *Otus rufescens* 21cm

This is the most likely of several scops-owls to be heard and seen, and the only one typical of agricultural and disturbed areas. Dull brown or greyish-brown, with pale buff half-collar on rear of neck; short ear tufts and striking buff eyebrows above the brown eyes. Below, warm buff with short streaks. Call a soft single *hooup* repeated about every 12 seconds; begins soon after dusk. Feeds mainly on insects, small rodents; perches in middle storey. Found in secondary vegetation, logged and unlogged forest, and oil-palm and rubber estates. Occurs through most of east Asia to the Philippines; resident.

BARRED EAGLE-OWL *Bubo sumatranus* 45cm

Though this is one of the smaller eagle owls worldwide, it is equal to the biggest of the owls of Borneo. A heavy owl, with brown eyes and yellowish bill set in pale face-mask; narrow black barring on the whitish underparts, becoming brown on breast; mottled brown above and on crown. Head often looks flattish because of elegant eyebrows. Juveniles very pale, almost white. Call a deep, double *hoo hoo*, and a harsh quacking *kakakakakak*. Found in logged and unlogged forest in lowlands, seldom far up hilly land, sometimes near rural villages. Occurs from south Burma to Java and Borneo; resident.

BUFFY FISH-OWL *Ketupa ketupu* 45cm

One of the biggest owls in open country; its size and the staring yellow eyes are good identification features. Rufous with some buff spotting and black streaks above, reddish-buff with narrow black streaking below; large ear tufts, eyes brilliant yellow. Both local and scientific names are based on wild, ululating four-note call; other calls also known. Feeds on fish, frogs, other small animals; footprints sometimes seen in sand at edge of water. Found in forest near rivers, tall secondary vegetation, agricultural estates. Occurs from west Burma to Indochina and Java; resident.

BROWN WOOD-OWL *Strix leptogrammica* 45cm

Because of its size alone, observers can sometimes jump to the conclusion that this is an eagle owl; it is important to note the face pattern. Dark brown eyes surrounded by dark smudges, set in deep rufous face-mask; underparts pale but finely and closely barred blackish (with heavier barring than in Peninsular Malaysian individuals), breast dusky. Upperparts mottled dark brown, no pale spots on crown. Short eyebrows and throat pale, buffish. Call four notes, the first the strongest; and variations on this theme. Found in lowland and hill forest. Occurs from India to Borneo and Java; resident.

BROWN BOOBOOK *Ninox scutulata* 30cm

The lack of a proper face-mask is characteristic of this owl. Plain dark brown head and upperparts, with rather long, broadly barred tail; whitish below with wide, heavy brown streaks; eyes yellow, with pale fluffy feathering at base of bill between eyes. Easily identified by call, a rapid repeated *kewick kewick kewick*, often for minutes on end. Found in unlogged and logged forest, mangroves, secondary vegetation, sometimes rubber or oil-palm estates. Occurs through most of east Asia, from India to Sulawesi; here resident in forest and mangroves, migrant in other habitats.

GOULD'S FROGMOUTH *Batrachostomus stellatus* 22cm

This is always the most commonly encountered frogmouth in the lowlands. Slightly larger than Javan, either bright rufous or dark brown above, breast grey-brown with tawny scaling and buff spots lacking any black margins; tail with doubled light and dark cross-bars. Female less greyish, darker and more chestnut than male. Sunda Frogmouth *B. cornutus* similar but (male) thickly vermiculated with black above and below, and tail with only simple dark cross-bars. Perched by day, flying low through understorey by night; call an eerie wheeze, *hoo-hooiu*. Found in lowland forest. Occurs in the Malay Peninsula, Sumatra and Borneo; resident.

BLYTH'S FROGMOUTH *Batrachostomus affinis* 19cm

Finding frogmouths requires some dedication, and identifying them needs great care; most have two colour phases. Male grey with rufous wash, finely camouflaged like tree-bark, no bold markings above, below or on head. Female rufous with (some, not many) bold black-edged white feathers on breast, and (some, not many) soft bristles round bill and ears. Of other frogmouths, Large *B. auritus* is twice as bulky, Gould's has scaly buff-spotted breast. Perches motionless on branch or stump by day, calling from perch (a descending series of *quarks*) by night. Found in forest, edge growth. Occurs from the Malay Peninsula to Borneo and Java; resident.

LARGE-TAILED NIGHTJAR *Caprimulgus macrurus* 30cm

Of two open-country nightjars this is the commoner in towns, with a loud call; the Savanna Nightjar *C. affinis* in Kalimantan gives unimpressive *churrs* in flight. Finely barred and mottled greyish-brown all over, with pale moustache, white throat patch, and broad ashy grey eyebrows separated by dark centre crown. Male has white patch on centre of primaries and white corners to tail; these are duskier, inconspicuous in female. Call a loud *klok*, like knocking on wood, slowly repeated many times, often in bouts of two or three with pauses. Found in open country, agricultural land, scrub, waste ground. Occurs throughout south Asia to Australia; resident.

GLOSSY SWIFTLET *Collocalia esculenta* 10cm

Unlike some other swiftlets, this species does not echo-locate, and therefore is not found deep inside dark caves. Very small, all black (glossy above when seen close to) except for dusky whitish abdomen; tail forked very slightly. No other swiftlet has pale belly. Hawks for small insects in flight, keeping close to vegetation surfaces, trees; nests in abandoned buildings, tunnels, light cave entrances. Found over forest, secondary growth, gardens and other habitats to 1,500m. Occurs through South-east Asia to Australia; resident.

HOUSE SWIFT *Apus nipalensis* 15cm

The most familiar of swifts in towns, they can form big flocks and typically nest in colonies. Black, with brown flight feathers, white rump and throat; tail with shallow fork, often spread so fork not seen. The combination of white throat and rump is shared only by Fork-tailed Swift *A. pacificus* which is bigger, with longer and more pointed wings, and longer and deeply forked tail. In flight, a shrill screaming or chattering; nesting on buildings, under bridges, cliffs. Found in all open habitats, including city centres to hilltops, coasts, offshore islands. Occurs from Nepal to the Philippines; recently split from Little Swift *A. affinis* of Africa and India; resident.

DIARD'S TROGON *Harpactes diardii* 30cm

One of six Bornean trogons, this species has (in the male) an odd and inconspicuous patch of violet on the crown. Male has black head, throat and upper breast, separated from scarlet breast and cinnamon-brown back by poorly visible pink line. Female has brown head, back and breast, dirty pink abdomen; in both sexes the tail is long, black above, largely white below with black speckles and vermiculations. Perches in lower and middle storey, taking short flights to new perches, turning head slowly in search of insects. Call four mournful and descending notes. Found in lowland forest. Occurs in the Malay Peninsula, Sumatra and Borneo; resident.

SCARLET-RUMPED TROGON *Harpactes duvaucelii* 24cm

The male's brilliant facial skin may be significant in courtship or in indicating the position of nest holes excavated by the male. Male has black head, a scarlet breast and abdomen, cinnamon-brown back and bright patch of scarlet on rump; no pink line bordering black of head; bright blue skin above eye, deeper blue bill. Female has dirty pink rump and upper tail-coverts; pink abdomen and under tail-coverts; in both sexes the tail is white below without speckles. Perches in lower and middle storey, turning head slowly to look for insects; call a rapidly descending series of about a dozen notes. Found in lowland forest. Occurs in the Malay Peninsula, Sumatra and Borneo; resident.

WHITEHEAD'S TROGON *Harpactes whiteheadi* 33cm

Male (above); female (right)

A spectacular montane bird, the only trogon in Borneo whose crown and throat are contrasting colours. Male with scarlet crown, blackish throat grading into grey, then sharp cut-off to scarlet abdomen; the back and rump cinnamon. Female similar but crown, back, lower breast and abdomen cinnamon, contrasting with blackish throat that grades into grey breast. In both sexes, backlighting makes the grey breast appear white; take care to distinguish Diard's and Red-naped Trogon *H. kasumba*. In Borneo, male Whitehead's is the only trogon with a red crown. Found in lower and upper montane forest above 1,000m, in middle strata, often singly. Occurs only in Borneo.

COMMON KINGFISHER *Alcedo atthis* 18cm

This is one of the most widespread kingfishers in the world, familiar throughout Europe and Asia. The smallest of open-country kingfishers, blue-green on crown, moustache, wings and back; shining blue stripe down back; rufous on ear coverts, breast and abdomen. Black bill, tiny red feet. Flight direct, low, showing blue back stripe. Call a simple high piping repeated two or three times, not useful for identification. Found at water's edge in open country swamps, old mining pools, aquaculture ponds, mangroves. Occurs throughout the Old World; absent from Australia; resident in lowlands and occasionally at higher altitudes.

BLUE-EARED KINGFISHER *Alcedo meninting* 15cm

A small, deep blue kingfisher, this bird can sometimes be seen darting from its nest in the earth banks of forest streams, and is the forest equivalent of the open-country Common Kingfisher. Like that species but deeper blue, with blue (not rufous) ear coverts, deeper rufous underparts, and a deeper but still brilliant iridescent blue line down back to rump best seen in flight. Bill black with some reddish colour at base, tiny feet red. Solitary, best seen at nest sites. Found in lowland forest up to 900m, commoner in the level lowlands, and maintaining a foothold in overgrown plantations near forest. Occurs from India to Java and the Philippines; resident.

STORK-BILLED KINGFISHER *Pelargopsis capensis* 37cm

The biggest kingfisher in Borneo, this species is not uncommon along the coast and large rivers, and shows a sky-blue stripe down the back in flight. Bill red; head brown becoming light rufous on breast and abdomen, and on hind neck; blue wings and brighter blue back; longish tail. In flight, big red bill and plain blue wings. Call a loud *kow-koo*, quite pleasant, or a harsher variant when these two notes extended into repeated cackle. Found near coasts, mangroves, agricultural land near sea including rice fields and coconut plantations; larger rivers where bordered by forest. Occurs from India to Sulawesi; resident.

BLACK-CAPPED KINGFISHER *Halcyon pileata* 30cm

A migrant to the region, this elegant kingfisher often appears in agricultural areas such as oil-palm and rubber estates, if there is water available. Red bill and feet; black crown and sides of head, sharply cut off from buffish-white collar and throat, becoming fulvous on abdomen; blue back and tail, black carpal joints. In flight purplish-blue with black wing-tips and white patch at base of primaries. Call a shrill laugh. Found mainly in freshwater habitats, forested rivers, reservoirs, swamps, also mangroves. Occurs throughout east Asia from India to Sulawesi; migrant.

COLLARED KINGFISHER *Todirhamphus chloris* 24cm

The sea-green plumage of this kingfisher is symbolic of its coastal habitat. Blackish to grey bill and feet; turquoise-blue crown, back, wings and tail, with white underparts and broad white collar bordered by narrow black line. In flight, wings plain greenish-blue, the tone of blue very variable according to the light. Call is a harsh series of *kek-kek, kek-kek*, notes given in couplets, harsh and high-pitched, usually when flying. Found in mangroves, coasts, sandy beaches, and occasionally a considerable distance inland in large gardens and agricultural estates. Occurs from the Middle East to the western Pacific; resident.

BLUE-THROATED BEE-EATER *Merops viridis* 27cm

Nesting in open country, many individuals move away into forested habitats during the non-breeding season. Bright chestnut crown and back; rest of plumage bright light green, blue on throat and tail with elongated central tail feathers. Distinguished from the Blue-tailed Bee-eater by chestnut head and back; no rufous on throat. Seen in small groups or singly, sallying out from perches on exposed twigs, wires. Found in open habitats from April to August, and in forest canopy and forest edge all year in small numbers. Occurs from south China to the Philippines, Sumatra, Java and Borneo; resident and partial migrant.

BLUE-TAILED BEE-EATER *Merops philippinus* 30cm

A bird of open country in Borneo during the northern winter, this bee-eater is seldom seen together with the previous species. Bright sage-green, paler below, shading into blue on rump and tail; black mask through eye, chin yellow and throat with a rufous patch. Two central tail feathers elongated into points. In flight, wings chestnut below; glides more than the previous species. Call a shrill *chiwi* in flight. Found in all open habitats, rural areas, mangroves, plantations, scrub and secondary growth. Occurs from India to New Guinea; migrant, some breeding records from north.

DOLLARBIRD *Eurystomus orientalis* 30cm

Its name is purportedly derived either from the silver coin-shaped markings on its wings, or from its tumbling flight tracing $ signs. Heavy, bright red bill, small red feet; a big-headed, dark bluish bird, often appearing blackish, perching erect against sky in treetops or on wires. Juvenile similar but duller, with blackish bill. Call a harsh *kek-kek*, often given in flight, when pale wing patches and occasional tumbling are seen. Found in open country, secondary growth, forest edge and mangroves. Occurs from Himalayas to Australia and New Zealand; resident, in coastal lowlands mainly.

COMMON HOOPOE *Upupa epops* 30cm

Familiar in picture and folklore throughout the drier subtropics of Asia, this bird's small size takes first-time observers by surprise. Light rufous-pink head, crest and underparts, pink upper back and carpal joints; crest feathers tipped black, and wings and tail boldly banded black and white. Often on the ground, probing for insects with thin curved bill; solitary or in small, loose groups. Call a dull booming, three or four notes. Found in open country, scrub, secondary growth, only in the north. Occurs throughout the Old World to South-east Asia; in Borneo an extremely rare vagrant.

WHITE-CROWNED HORNBILL *Berenicornis comatus* 100cm

This unkempt hornbill can occur low down in the forest understorey, and has even been seen on the ground. Dull greyish horn-coloured bill, ragged puffy white head, white underparts, tail and wing tips; back and wings black. Male has white breast, female black. White parts of plumage often stained and untidy. In flight, black with white trailing edge to wing, white head, breast and tail. Call a repeated three-note *huh hoo, hoo*, the first note least emphatic. Found in lowland forest, in the middle or even the lower storey. Occurs from Indochina to Sumatra and Borneo; resident.

BUSHY-CRESTED HORNBILL *Anorrhinus galeritus* 88cm

Keeping in contact by their continual seagull-like yapping, family parties co-operate in feeding the breeding female of this species on her nest. Greyish-black, the tail paler grey with a broad black tip; inconspicuous bunch-shaped crest, and bare bluish skin on face and throat. Bill and feet blackish; the least colourful hornbill. Found in lowland and hill forest up to 1,200m altitude, keeping to middle and upper storey. Occurs in the Malay Peninsula, Sumatra and Borneo; resident.

WRINKLED HORNBILL *Aceros corrugatus* 85cm

Male (left); juvenile (above)

Care is needed in looking methodically at the plumage to distinguish this bird from the Wreathed Hornbill. Black, with base of tail black and the rest white stained yellow. Male has white head and neck with short black crest, white throat pouch, reddish casque on top of bill and base of bill. Female has black head and neck, blue throat pouch and skin round eye. Thinly distributed, never in big flocks; call a bark often of two notes. Found in coastal and lowland forest, occasionally in hill forest. Occurs in the Malay Peninsula, Sumatra, Borneo; resident.

WREATHED HORNBILL *Aceros undulatus* 100cm

Perhaps still the commonest and most widespread of hornbills, this bird's rather simple call draws less attention than that of the bigger but rarer species. Black, with entire tail white or stained yellow. Male has a short chestnut crest and yellow throat pouch with black bar sometimes visible; little or no casque on bill. Female has black head and neck, blue throat pouch with black bar, but a little red skin round eye. Call a bark, one or sometimes two notes. Found in lowland and hill forest, common up to 1,300m. Occurs from east India and Burma to Borneo, Sumatra, Java; a resident but undertakes long-range movements.

BLACK HORNBILL *Anthracoceros malayanus* 75cm

Never found in large flocks, pairs of this small hornbill may also be accompanied by their grown young. Male is all black with a whitish horn-coloured bill and white corners to tail, only some individuals have a broad, dull whitish eyebrow. Female all black with blackish bill and some reddish facial skin. Best identified by call, a disgusting retching sound, irregular. In pairs or family parties, in middle storey. Found in lowland forest, not so conspicuous or vocal as other hornbills, but the species that survives best in forest fragments. Occurs in the Malay Peninsula, Sumatra, Borneo; resident.

71

ORIENTAL PIED HORNBILL *Anthracoceros albirostris* 68cm

Regional plumage differences are overshadowed by the variability among individuals of both sexes in features such as bill colour and tail-feather markings. Black, with both white abdomen and flanks, white wing-tips are seen as entire white trailing edge to wing in flight, and variable amount of white on lateral tail feathers. Shows whitish patches on face, pale horny bill with casque marked with black. Call is a clattering laugh. Found in forest edge, riverine forest, secondary growth in lowlands, especially coastal and on islands. Occurs from India to Sumatra, Borneo, Java; resident.

RHINOCEROS HORNBILL *Buceros rhinoceros* 90–120cm

Casque and feathers of this hornbill are motifs often used in Borneo art and dance. Black, with white abdomen and tail crossed by wide black band; bill yellow with red patch at base, the casque orange-red with a yellow tip. In flight Great and Helmeted Hornbills are the only others with black band on white tail, but Rhinoceros Hornbill has wings entirely black. Call *kronk*, differently pitched in male and female, duetting in flight *kronk krank, kronk krank*. Found in forest from lowlands to 1,300m. Occurs in the Malay Peninsula, Sumatra, Borneo, Java.

GOLD-WHISKERED BARBET *Megalaima chrysopogon* 30cm

This, the biggest of the barbets, is commonly heard but the bright colours are hard to see in the forest. Big; green with large yellow cheek patches, front half of crown yellow, and rear half of crown varying from red to red-flecked blue; dull grey throat. Call a loud repeated *ku-took*, more than one couplet per second, often for minutes on end. Also a repeated trill of similar tone, becoming briefer, slower until each trill has only three or four notes. Found in forest, forest edge, plantations, scattered woodlands, occasionally well-established old rural gardens with big trees. Occurs in the Malay Peninsula, Sumatra, Borneo; resident.

GOLDEN-NAPED BARBET *Megalaima pulcherrima* 20cm

This is one of the three barbets endemic to Borneo, all of them montane, and the only one of the green barbets to lack any red markings at all on the head. Overall green plumage, with crown, chin and throat bright turquoise-blue, but not sharply defined and in poor light appearing same colour as body; nape suffused with golden-yellow, again poorly defined. Call a four-part *took took tu-rook*, repeated but not very persistent; also a repeated trill becoming briefer, like Gold-whiskered. Found in canopy of tall montane forest, forest edge, about 1,000–2,000m. Occurs only in Borneo; resident.

73

SPECKLED PICULET *Picumnus innominatus* 10cm

This is a characteristic montane forest bird, sometimes rather hard to find, which lacks bright colouring in either sex. Very small with bright olive back, blackish head bearing two white stripes on brow and moustache; underparts speckled black on white. Male has yellowish patch on forehead, dark grey in the female (shown). Forages on trunks and the branches of small trees, seldom in the high canopy but often near forest edge, singly or pairs, drumming or giving short sharp call, *tsick*. Usually seen alone, but sometimes in mixed species flocks. Found in montane forest, at about 900–1,500m. Occurs from the Himalayas to Borneo; resident.

SUNDA PYGMY WOODPECKER *Dendrocopos moluccensis* 15cm

Of two similar-looking little woodpeckers in this genus, this one is a specialist in mangrove and coastal habitats, whilst the other occurs in the canopy of inland forest. Small woodpecker with brown crown, white barring on dark brown back and wings; broad blackish moustache stripe and white throat, and whitish streaked underparts. Male has small red streak on side of crown behind eye. Seen singly or in pairs, feeding low on small trees; call a vibrating trill, not striking. Found in mangroves and secondary growth near coast, and occasionally inland. Occurs discontinuously from India to Borneo and Java; resident.

BANDED WOODPECKER *Chrysophlegma mineaceus* 25cm

Several woodpeckers in the region have bright-yellow-tipped crests, and red and green plumage, making them seemingly easy to identify, but the important features to look for are not the most conspicuous ones. Male has red head with bright yellow-tipped crest; upperparts dull greenish, finely banded; underparts dull maroon strongly banded with buff; wings maroon, some banding on primaries. Female similar but sides of head dull. Barring above and below separate it from Crimson-winged Woodpecker. Found in lowland forest, plantations, mangroves, scattered trees in rural areas. Occurs in the Malay Peninsula, Sumatra, Borneo and Java; resident.

CRIMSON-WINGED WOODPECKER *Picus puniceus* 25cm

Like the previous species, this fine woodpecker is found not only in forest but occasionally in plantations and even gardens. Dark olive-green above and below with plain maroon-red wings; no barring above or on wings, little barring beneath on flanks, sometimes to abdomen and under tail-coverts. Red crown and crest with yellow tip; male has red moustache stripe. Juvenile more barred, may be confused with Banded Woodpecker but has green head. Call a two-note *chee chee*, emphatic first note. Found in lowland forest, secondary vegetation, plantations. Occurs in the Malay Peninsula, Sumatra, Borneo and Java; resident.

CHECKER-THROATED WOODPECKER
Chrysophlegma mentalis 28cm

Larger than the Banded and Crimson-winged Woodpeckers, this species is more confined to virgin forests. Chestnut collar separating olive-green head from green body; no red on head but a black and white chequered patch on throat; wings foxy red, with narrow black cross-bars on primaries. Distinguished from Banded and Crimson-winged Woodpeckers by lack of red on crown, and by throat pattern. The chequered throat reaches to the cheeks in male, where it is replaced by chestnut in female. Found in lowland, hill dipterocarp and lower montane forest, up to 1,200–1,600m limit. Occurs from Peninsular Thailand to Sumatra, Borneo, Java; resident.

COMMON FLAMEBACK *Dinopium javanense* 30cm

Although this species is similar in appearance, it is not closely related to the Greater Flameback *Chrysocolaptes lucidus*. It is golden-brown on the back and wings, has a bright red rump and black flight feathers; scalloped black on white underparts. Head boldly striped black and white, with single black moustache stripe (double in Greater Flame-back); male has prominent scarlet crest, female crestless with black crown streaked white (spotted white in Greater Flameback). Found in: plantations; coconut; scattered woodland in lowlands. Occurs from India to the Philippines; resident.

MAROON WOODPECKER *Blythipicus rubiginosus* 22cm

Mainly found in undisturbed forest, this bird's deep colouring makes the pale bill stand out conspicuously. Generally looks very dark, plain maroon-brown without any barring, maroon on back and wings; male has scarlet patch round sides and back of neck. Small size and ivory-yellow bill. Usually alone, call a repetitive squeak when foraging, or a high-pitched descending trill. Found in understorey of primary forest, in extensive forest areas, and amongst bamboo in lowlands; has rarely been recorded in overgrown plantations. Occurs in the Malay Peninsula, Sumatra and Borneo; resident.

GREAT SLATY WOODPECKER *Mulleripicus pulverulentus* 50cm

A fine, noisy woodpecker, parties of which tend to stay high up in the forest canopy. Large and long-bodied, plain dark grey plumage and peachy buff throat; grey bill and legs. Male has small red patch in region of moustache. Nearest in size is the White-bellied Woodpecker *Dryocopus javensis*, black with white abdomen. Found in small groups; cackling laughter as they follow one another through canopy, noisy pecking and drumming on boughs. Found in lowland forest, to about 1,000m. Occurs in the Malay Peninsula, Java, Sumatra and Borneo; resident.

BLACK-AND-RED BROADBILL
Cymbirhynchus macrorhynchos 25cm

The nest of this brightly coloured bird is an untidy bundle of dead leaves draped from a branch or bamboo, looking like litter stranded there by floods. A chunky bird with black upperparts, red rump (concealed when perched) and deep red underparts crossed by a black breast-band; long white streak down scapulars. Bill brilliant cobalt-blue above and yellow below. Juvenile similar, dull reddish-buff underparts. Call a variety of harsh chucks and wheezes. Found in lowland forest, overgrown plantations, and tall secondary growth, especially along rivers. Occurs from Burma to Indochina, Borneo; resident.

BLACK-AND-YELLOW BROADBILL *Eurylaimus ochromalus* 16cm

The characteristic call of this bird in the forest is sometimes triggered by loud noises such as thunder, falling trees, or the calls of other animals. Small, stubby, black and white; black head with complete white collar bordered below by black; pink breast, yellow rump and yellow flashes on black wings. Brilliant turquoise and yellow bill, yellow eye. Call a long trill starting slowly *dee, dee, de de de...*, without the introductory bang heard in call of Banded Broadbill *E. javanicus*. Found in lowland forest to 700m, logged forest, overgrown plantations. Occurs in the Malay Peninsula, Sumatra and Borneo; resident.

LONG-TAILED BROADBILL *Psarisomus dalhousiae* 28cm

Small groups of this bright parrot-like bird occur in the montane forest, keeping mainly to the higher trees. Adult brilliant green with yellow face and throat, black cap showing blue crown and nape, lime-coloured ear tufts. Tail long, graduated, blue; wings black with large blue patches that appear white from below. Bill bright yellowish-green. Juvenile duller, with head green. Call five or six screeches all of same tone. Found in montane forest about 800–2,000m, in the canopy and middle storey. Occurs from the Himalayas to Sumatra, Borneo; resident.

GREEN BROADBILL *Calyptomena viridis* 18cm

The green species of broadbill have dense feathering and plush head plumages that almost conceal the bill; in other broadbills the bill itself is brilliantly coloured. Chunky birds, green, the male almost glowing green with three sharply defined black marks on folded wing, black marks before and behind eye. Female paler green, without the black marks. Call a descending series of short pops, accelerating, like a ping-pong ball dropped on table but melodious. Found in lowland forest, in middle and lower storey, often near rivulets. Occurs in the Malay Peninsula, Sumatra, Borneo; resident.

Male (above); female (below)

WHITEHEAD'S BROADBILL *Calyptomena whiteheadi* 25cm

Male (above); female (below

Much the biggest of the three green species of broadbill in Borneo, this is one of the most magnificent endemic species. Male an intense iridescent green, the dark bases of feathers showing through in places to give a chainmail effect; black throat, ear coverts, black bands on wing coverts, black scales on crown. Female is a less intense apple-green, fewer black markings but retaining black throat. Seen in small parties or singly in middle storey, usually rather quiet but can be noisy in groups, short harsh shrieks. Found in tall lower montane forest in valleys, about 1,000– 1,500m. Occurs only in Borneo; resident.

BANDED PITTA *Pitta guajana* 23cm

Good light conditions are needed to see this magnificent bird at its best, when the blue and orange colouring seems iridescent. Crown and sides of face black with very broad orange brows; throat white, back brown, darker in the male. Underparts in male deep blue with orange bars across sides of breast; in female golden-buff with fine black crossbars. Wings dark with broad white tips on wing coverts. Juvenile light buffish brown with buff eyebrows, bluish tail, white wing coverts. Call a short *trrr* falling in tone, and a falling whistle *pow*. Found in lowland forest. Occurs in the Malay Peninsula, Sumatra, Borneo, Java; resident.

80

BLUE-HEADED PITTA *Pitta baudii* 18cm

In this pretty Borneo endemic there is striking dimorphism between the sexes. Male foxy chestnut above; crown, tail and underparts bright blue, almost iridescent on the head; black mask through eye, and the throat and two wingbars white. Female rufous above and fawn below, with blue tail, pale throat and two whitish wingbars. Hopping along ground, seeking insects amongst leaf litter; usually solitary but fairly common in suitable habitat. Found in extreme lowland forest near rivers, lowland forest generally. Occurs only in Borneo; resident.

BLUE-WINGED PITTA *Pitta moluccensis* 20cm

From the more northerly parts of Asia this pitta migrates in small numbers, not in flocks, and at that time can be found lurking in a wide range of habitats. Black head with striking buff stripes at side of crown, white throat and fulvous breast and flanks; bright scarlet abdomen. Back green, grading into blue rump; wings with blue carpal joint, sometimes white visible on primaries. In flight, big white patch on primaries, wings otherwise dark with blue base. Distinguished from the very rare Mangrove Pitta *P. megarhyncha* by smaller, paler, clearer crown stripes, slower call. Call a two-note whistle with emphasis on second note. Found in lowland forest, tall secondary growth, other habitats when migrating. Occurs from India to the Philippines; migrant.

81

HOUSE SWALLOW *Hirundo tahitica* 14cm

A widespread swallow found throughout the year, whereas the northern Barn Swallow *H. rustica* is a migrant. Distinguished from the latter by smaller size, sullied greyish underparts with no dark band bordering rufous throat; rufous extends onto forehead. Dark under tail-coverts are marked with white, tail forked but outer feathers not long. Juvenile browner above, less rufous on forehead and throat. Found in all open-country habitats, over mangroves and lowland forest, to above 2,000m, breeding under bridges and various other man-made structures. Occurs from India and Taiwan southwards through to New Guinea and west Pacific; resident.

STRIATED SWALLOW *Cecropis striolata* 18cm

Heavier in build than other swallows, Striated Swallows (now split from the Red-rumped Swallow *C. daurica*) arrive in Borneo from various parts of Asia and differ somewhat in markings. Crown, back and tail dark with blue iridescence; deeply forked tail; chestnut rump, collar on hind neck and occasionally breast, entire underparts heavily streaked blackish on cream background; the chestnut varies greatly. Seen in small numbers, in rural areas especially near limestone outcrops. Found in lowlands, feeding over agricultural land and secondary growth. Occurs from Africa through southern Eurasia to the Philippines; in Borneo a regular migrant in small numbers.

EASTERN YELLOW WAGTAIL *Motacilla tschutschensis* 17cm

Many of the juvenile and non-breeding birds occurring in Borneo are very pale, with almost no yellow in the plumage. Adult olive brown above, yellow to cream below, with white outer tail feathers; head darker olive with pale eyebrow. Juvenile similar but light-brownish olive above and white to buff below, usually with a few dark specks on the breast; eyebrow nearly white. Found on short grassland, coastal scrub and bare ground, sometimes in ricefields up to high altitudes, foraging for insects on bare ground and turf. Occurs in Eastern Asia, recent split from Yellow Wagtail *M.flava*.

PADDYFIELD PIPIT *Anthus rufulus* 17cm

The migrant and resident forms of what was once known as Richard's Pipit *A. richardi* are increasingly being recognized as distinct species; resident birds are now split as the Paddyfield Pipit. Not always told apart in the field, both are warm brown streaked with dark brown above, buff below with faint breast streaks, and a prominent buff eyebrow; white outer tail feathers. Paddyfield Pipits smaller, with less distinct breast streaks. Pink legs; trots and pauses on roadside verges, short grass, open agricultural land. Occurs throughout South-east Asia to Philippines.

PIED TRILLER *Lalage nigra* 17cm

Together with cuckoo-shrikes and minivets, Pied Trillers form a diverse family which is at its most diverse in this region. Male black and white with a rather flat-headed profile, striking white eyebrow and face, white wingbar; black above and on white-sided tail, with a large grey rump patch. Rump patch and wingbar visible at rest distinguish this species from Ashy Minivet. Female brownish-grey above, paler grey below, with striking white wingbar; faint pale eyebrow; underparts and rump finely barred. Found in gardens, plantations, secondary growth and mangroves. Occurs from the Malay Peninsula to Sulawesi and the Philippines; resident.

ASHY MINIVET *Pericrocotus divaricatus* 20cm

The crisp, clean appearance and long white brow help to distinguish this visiting minivet from the Pied Triller; it perches upright, and forms flocks. Male has bright white face including forehead, dark crown and dark line from bill through eye; underparts white to very pale grey, back and wings plain grey but revealing a white bar across wing in flight. Tail dark, the outer feathers mainly white. Female paler grey on crown, less white on forehead. Found in forest, forest edge, riverside and secondary vegetation. Occurs from China to the Philippines, Borneo and Sumatra; migrant October to April.

GREY-CHINNED MINIVET *Pericrocotus solaris* 18cm

Male (above); female (below)

This montane bird can be separated from the Scarlet Minivet *P. speciosus* by carefully examining the wing pattern, which is not easily visible in canopy-living birds. Male has grey throat and face that need to be looked for carefully; underparts more orange but juvenile male Scarlet is also somewhat orange; no separate patch of colour on secondaries. Female grey and yellow as other minivets but throat whitish (not yellow) and no trace of yellow on entirely grey forehead. Behaviour similar to other minivets, in small parties or more often pairs. Found in hill and montane forest, about 800m upwards. Occurs from the Himalayas to Indochina and Borneo; resident.

BAR-WINGED FLYCATCHER-SHRIKE *Hemipus picatus* 13cm

This is a fairly common small black-and-white bird that is often seen in mixed foraging flocks with babblers or bulbuls. Male black above and white below, with white rump and outer tail feathers; white wing panel distinguishes it from the Black-winged Flycatcher Shrike *H. hirundinaceus*, and lack of white eyebrow from Ashy Minivet and Pied Triller. Female like male but dark parts of plumage browner. Seen in small groups, seldom singly, in middle and upper storey, snatching insects. Found in hill and montane forest, 500–2,000m. Occurs from India to Borneo; resident.

STRAW-HEADED BULBUL *Pycnonotus zeylanicus* 28cm

The fine song of this big bulbul is often heard during river journeys in Borneo, and is formed by the duetting pair. Identified by size and by the rich rufous-buff crown that appears streaked or furrowed, with blackish moustache and mark through eye, pale throat and underparts. The breast and back have obscure narrow light streaks. Typical of, but not confined to, riverbanks, where the lilting song is richer, more melodious than that of any other bird. Found in forest and forest edge, along rivers, sometimes surviving in small forest remnants, and plantations in the lowlands. Occurs in the Malay Peninsula, Sumatra, Borneo and Java; resident.

BLACK-AND-WHITE BULBUL *Pycnonotus melanoleucos* 18cm

This seldom-seen bulbul is thought to be semi-nomadic, presumably following the patterns of fruit production of different forest trees. Adult black with a rounded white patch on wing; in flight, white beneath wing. Juvenile dark olive-brown, rump slightly paler rufous; intermediates acquiring black plumage occur. Habits, shape and ragged edges of wing patch tell it from black race of Magpie Robin in east Borneo. Seen singly, seldom in mixed foraging flocks, feeding on small fruits high up in canopy. Found in lowland and hill forest. Occurs in the Malay Peninsula, Sumatra and Borneo; resident.

BLACK-HEADED BULBUL *Pycnonotus atriceps* 17cm

This is a sociable bulbul, brightly coloured with an enamel-blue eye. Rather small, with black head and throat, and greenish-olive above and yellow below; olive tail with yellow tip and black subterminal bar. Tail pattern seen well in flight. Close up, iridescence on head and the bright blue eye are very attractive. It is typically seen in small parties, individuals following each other from bush to bush and calling with loud single notes, tui. Found in forest and the forest edge, secondary growth, low scrub along rivers and roadsides up to 1,000m. Occurs from east India to Indochina, Borneo, Java; resident.

BORNEAN BULBUL *Pycnonotus montis* 18cm

The unusual pointed crest is enough for reliable separation from all other bulbuls in Borneo. Tall black crest stands erect on black head, with contrasting yellow throat; olive back and tail, and bright yellowish underparts. Recently split from mainland Black-crested Bulbul *P. melanicterus* (which is similar but has a black throat) and now regarded as a Bornean endemic. Separated from Black-headed Bulbul by its crest, pale eye, and plain tail. Call a hesitant, repetitive song of five or six notes, the last two repeated. Found in lowland and hill forest, forest edge and tall secondary growth from lowlands to 1,200m. Endemic resident.

SCALY-BREASTED BULBUL *Pycnonotus squamatus* 14cm

This easily identified bulbul tends to be found in hilly country, keeping to the tree-tops and seldom mixing with other species. Small, with black head and contrasting white throat, and black-and-white scaled breast; distinctive yellow under tail-coverts; brown above, with white tips to outer tail feathers seen in flight. Often solitary, feeding on small fruits in canopy. Call a series of high single notes, not very distinctive. Found in forest in lowlands and hills. Occurs in the Malay Peninsula, Sumatra, Borneo and Java; resident.

PUFF-BACKED BULBUL *Pycnonotus eutilotus* 22cm

This bulbul is under-recorded because of its nondescript appearance, which makes it hard to distinguish from three or four other brown bulbuls. Large, plain brown above and whitish below with greyish tinge to breast and buff tinge to abdomen. Tips of outer tail feathers typically but not always whitish, hard to see; eye reddish-brown. There is a faint crest, giving a slightly peaked profile to hind crown. Call four notes, the first emphatic and the remainder rapid, running together. Found in forest and forest edge, keeping to middle and lower storey. Occurs in the Malay Peninsula, Sumatra and Borneo; resident.

PALE-FACED BULBUL *Pycnonotus leucops* 19cm

Superficially similar to Yellow-vented Bulbul, in practice this species is usually not difficult to identify through a combination of plumage and habitat. Greyish-brown on crown, back and wings, the face is rather bright whitish, without a black line through the eye; breast off-white, the under tail-coverts bright yellow. Usually seen singly or in pairs; a simple song given from tops of bushes or low trees. Found in montane forest and forest edge, at 1,000m to above 3,000m. Endemic, formerly a race of the mainland Flavescent Bulbul *P. flavescens*. Resident.

YELLOW-VENTED BULBUL *Pycnonotus goiavier* 20cm

It is impossible to avoid seeing this, the commonest of all bulbuls and one of the best-known garden birds. Brown above, the head chalky white with a dark crown and dark stripe through eye; underparts white, tinged brownish-grey on breast which looks faintly mottled; under tail-coverts pale yellow. Crest often raised into short peak. Call a brief bubbling series of notes, often from wires or low trees, the two birds of a pair calling and raising spread wings over their backs in unison. Found in gardens, plantations and other cultivated land, towns, open country of all sorts up to hill stations, and in mangroves. Occurs from south Burma to the Philippines and Java; resident.

OLIVE-WINGED BULBUL *Pycnonotus plumosus* 20cm

This is one of about five rather similar-looking brown bulbuls, in which eye colour and slight differences in the tone of the breast, abdomen and wings are important. Eye dull red in adults, brown in juveniles. Plumage dark brown above, greyish-buff below with deep buff under tail-coverts. Ear coverts have faint whitish streaks, and feathers of folded wing have narrow olive fringes which together form inconspicuous olive panel. No crest, and larger than most other brown bulbuls. Found in secondary growth and scrub, along logging tracks, keeping to lower storey. Occurs in the Malay Peninsula, Sumatra, Borneo and Java; resident.

RED-EYED BULBUL *Pycnonotus brunneus* 18cm

Another of the confusing brown bulbuls, this one is distinguished from three other common species by red eyes that lack a striking yellow rim. Intermediate in size between Olive-winged and other brown bulbuls; dark brown above, brownish throat and breast with little contrast, with slightly paler abdomen and under tail-coverts. No distinguishing wing panel, but wings and tail may look faintly washed with olive. Calls and behaviour not distinctive; a high-pitched trill ascending at the end. Found in lowland forest up to 900m, secondary growth along forest edge. Occurs in the Malay Peninsula, Sumatra and Borneo; resident.

SPECTACLED BULBUL *Pycnonotus erythropthalmos* 18cm

This is one of the more attractive and easily identified of the brown bulbuls, by sight or sound. Brown above, pale below, with white throat and grey-tinged breast; eye is red or reddish-brown with narrow but conspicuous contrasting yellow-orange ring of skin round eye. Seen singly, or in small parties, often in mixed feeding flocks with other bulbuls and variety of bird species. Call is a repetitive three-note trickle, *diddle-ee, diddle-ee*. Found in lowland forest, forest edge, along logging tracks, tall secondary forest and peat swamp forest to about 1,000m. Occurs in the Malay Peninsula, Sumatra and Borneo; resident.

OCHRACEOUS BULBUL *Alophoixus ochraceus* 22cm

Primarily montane, this bird also inhabits hill slopes and overlaps with the similar Ashy Bulbul. Crown feathers often erected into moderately long peaked crest, and striking white throat is often puffed out. Eyes bright reddish-brown; greyish face, grey-brown back and buffish-brown breast grading into buff abdomen and fulvous under tail-coverts. Distinguished from Ashy Bulbul by no dark face, browner breast, from Grey-cheeked Bulbul *A. bres* by less yellow below. Call an unpleasant, harsh little song, repeated by and catching on amongst members of small flocks that move through middle storey. Found in hill and montane forest, 600–1,500m. Occurs from Thailand and Indochina to Sumatra and Borneo; resident.

91

YELLOW-BELLIED BULBUL *Alophoixus phaeocephalus* 20cm

This is a brightly coloured bulbul of the middle and lower storey that is often glimpsed flying rapidly and straight through the vegetation while giving harsh alarm calls. Grey head and upper breast, with white throat often puffed out, olive back grading into brown wings and tail; lower breast and abdomen bright pale yellow; tip of tail yellow in Sarawak and Kalimantan, not in Sabah. Seen in pairs and often with mixed foraging flocks of other birds, in lower to middle storey. Found in lowland and hill forest to about 1,000m. Occurs in the Malay Peninsula, Sumatra and Borneo; resident.

BUFF-VENTED BULBUL *Iole olivacea* 19cm

This is not a well-known bird, and its head pattern is more reminiscent of some babblers than of bulbuls. Has an overall creamy brown appearance, washed with grey on throat and upper breast, washed with cream on lower breast and abdomen, to buff under tail-coverts; centre of crown rufous, separated from pale brows by a darker line, little or no crest. Eyes dirty grey-brown; combination of eye and crown colour distinctive. Sociable, in groups especially foraging flocks with other species; noisy. Found in lowland and hill forest, forest edge, tall secondary growth. Occurs from south Burma to Borneo and Palawan; resident.

HAIRY-BACKED BULBUL *Tricholestes criniger* 16cm

Although its common name misleadingly points to a feature which is virtually impossible to see in wild birds, this bulbul is easily identified. Small; warm brown above, dirty yellowish below becoming yellower on abdomen; face with broad yellowish patch of feathers round eye and pale throat giving unique face pattern. A few very fine, dark hair-like feathers lie over the back. Common in middle and lower storey, usually solitary, occasionally in mixed feeding flocks. Found in lowland and hill forest, forest edge up to about 900m. Occurs in the Malay Peninsula, Sumatra and Borneo; resident.

STREAKED BULBUL *Ixos malaccensis* 22cm

Reasonably common along roads in forested hilly country, this bulbul is submontane rather than truly montane. Rounded crown without distinct crest; faint, narrow whitish streaks on grey throat and upper breast, white abdomen and white under tail-coverts; upperparts dark olive-brown not grading into brighter olive on wings or tail. Call a not very distinctive series of short trills; keeping more to tall forest rather than secondary growth or stunted elfin forest. Found in hill and lower montane forest about 500–1,000m altitude. Occurs in the Malay Peninsula, Sumatra and Borneo; resident.

CINEREOUS BULBUL *Hemixos cinereus* 20cm

Previously treated as a race of Ashy Bulbul *H. flavala* of the Malay Peninsula and Sumatra, Cinereous Bulbul is now recognized as a full species. It is less distinctively marked and needs greater care in identification. A striking white throat, often puffed out like that of Ochraceous Bulbul; dark crown and blackish smudge through and below eye; upperparts ashy grey including wings and tail, underparts nearly white. Call a varied and melodious song, compared with harsher, brief song of Ochraceous Bulbul; moving in small flocks in middle and upper storey. Found in hill and lower montane forest about 400–1,000m, degraded forest and forest edge. Resident. Endemic.

COMMON IORA *Aegithina tiphia* 15cm

The best views of this bird are when it is seen singing in bright sunshine, from the crown of a roadside tree. Adult yellow below, varying from olive-green to nearly black above; wings black crossed by two white bars; tail dark and unmarked. Female is slightly duller greenish-yellow, similar pattern. Rump may seem white, especially in flight, owing to the long white flank plumes. Call a series of quick, deep notes forming a slow trill; also a long descending whistle and a double descending whistle. Found in gardens, secondary growth, plantations, riverside trees, mangroves, forest edge. Occurs from India to Borneo and Java; resident.

GREEN IORA *Aegithina viridissima* 13cm

A delicately built bird of the forest, similar to leafbirds, but a deeper green and with wingbars. Male dark olive-green, the wings black with two white to cream wingbars and yellow edges to flight feathers; yellow spot above and below the eye forming nearly complete eye-ring, dark legs and bill. Female similar but duller, yellower, especially on throat. Call a long trill, ending in a powerful peow! Found in lowlands, primary forest, forest edge and tall secondary growth. Occurs in the Malay Peninsula, Sumatra and Borneo; resident.

GREATER GREEN LEAFBIRD *Chloropsis sonnerati* 20cm

This species has the best song of the leafbirds in the region, enriched by its ability to mimic other birds. Slightly larger than the other leafbirds, with a robust bill. Male has elongated black throat patch without a yellow border. Female has yellow throat and ring round eye. In flight and occasionally when seen perched, blue feathering on carpal joint is visible, but much less than in the Blue-winged. Call a varied song reminiscent of the Magpie Robin. Found in the middle storey of forest, forest edge, tall secondary growth up to 1,200m. Occurs in the Malay Peninsula, Sumatra, Borneo and Java; resident.

Male (above); female (below)

LESSER GREEN LEAFBIRD *Chloropsis cyanopogon* 17cm

Their feeding habits make it possible to watch leafbirds at forested roadsides, giving time to look for the necessary features. The Lesser Green is rather small, without blue on wing. Male has round black throat patch with narrow yellowish border. Female has green throat and no blue in wing. Call rather varied, a four-note whistle with third note the highest; feeding in middle storey and low trees, often in pairs feeding on small fruits. Found in forest edge in lowlands to 900m altitude, secondary growth. Occurs in the Malay Peninsula, Sumatra and Borneo; resident.

BLUE-WINGED LEAFBIRD *Chloropsis cochinchinensis* 17cm

The three species and two sexes of leafbirds in the lowlands – this, the Greater Green and Lesser Green – make a total of six hard-to-distinguish colour combinations. Both sexes of Blue-winged have turquoise-blue extending from carpal joint to primaries of wing, and less obviously on side of tail. Male has elongated black throat patch with distinct yellow border and overall yellowish tone to head. Female has green throat, like Lesser Green. Feeds on small fruits in the crowns of lower trees. Found in lowland and hill forest and forest edge up to 1,200m. Occurs from India to Borneo and Java; resident.

ASIAN FAIRY-BLUEBIRD *Irena puella* 25cm

While common, this magnificent bird is overlooked to a certain extent because it favours living high in the canopy. Male has shining enamelled blue upperparts including rump; black face, underparts, wing and tail; eye deep red. Female plain dark turquoise-blue with dark flight feathers, red eye. Calls include piping pee-dit, single notes, and frequent liquid song; singly, in pairs or small parties, moving from tree to tree. Found in lowland and hill forest up to 1,400m, in upper and middle storey especially at fruiting figs. Occurs from India to Borneo, Palawan, Java; resident.

TIGER SHRIKE *Lanius tigrinus* 18cm

This stocky, heavy-billed shrike is found in denser vegetation than the other species, especially in forest edge. Adult has grey crown and nape, rufous-brown back, wings and tail with narrow, wavy black bars, black mask through eye, and creamy white underparts. Juvenile like Brown Shrike but face mask dusky, barred, and extensive barring on upper- and underparts; thick bill. Seldom calls, and does not characteristically perch in exposed positions. Found mainly in lowlands, in forest, forest edge, secondary vegetation, overgrown plantations. Occurs through east Asia east to the Philippines and Sulawesi; migrant.

BROWN SHRIKE *Lanius cristatus* 19cm

Every year this common shrike arrives in great numbers within a very few days in mid Sept ember, remaining till April. Caramel or rufous-brown with black mask through eye, emphasized by pale forehead and eyebrow; more rufous on crown and tail, light buff to whitish below. Juvenile has faint and narrow blackish barring on flanks and back. Call a harsh chatter, especially in the early morning, mainly for a few weeks after arrival when setting up territories; perches erect on exposed twigs, fences, wires, catching insects on ground. Found in open country, cultivation, gardens, secondary growth. Occurs throughout east Asia to the Philippines and New Guinea; resident.

LONG-TAILED SHRIKE *Lanius schach* 25cm

Found only in south and east Borneo, this is an open-country bird, rather richly coloured and with a long tail. Black crown, nape and sides of face, plain bright rufous back, black wings with small rounded white patch at base of primaries; very long dark tail. Cheeks, throat, centre of breast and the abdomen are white, with rufous flanks, under tail-coverts and rump. Juvenile is similar but duller with some barring on flanks. Found in rice fields, open country including parks and canal margins, scrub, grassland, mainly in lowlands. Occurs from India to Taiwan, the Philippines and New Guinea; resident.

ORIENTAL MAGPIE-ROBIN *Copsychus saularis* 20cm

Male (above); female (right)

A familiar garden bird, this takes up the role of thrushes or robins elsewhere in the world. Male distinctively glossy black and white (nearly all-black in males in Sabah); head, breast, back, wings and tail black; white wing stripe, outer tail feathers and abdomen. Female similar but dark grey and white, juvenile more mottled than female. Hops on ground with raised tail, and in trees and bushes; a fine varied song. Found in all open country, cultivation, gardens, secondary forest, extending into disturbed forest and up rivers. Occurs from Pakistan to the Philippines, Borneo and Java; resident.

WHITE-RUMPED SHAMA *Copsychus malabaricus* 20–28cm

One of the more familiar birds of the forest understorey, this bird draws attention to itself by its calls. The head, breast and wings entirely glossy black in male, bold white rump and long graduated black tail with white edges; lower breast and abdomen rufous-orange. Female similar but dark grey and rufous rather than black and orange; shorter tail. Song commonly heard, and single angry chack notes are given from the understorey. Found in lowland and hill forest up to 1,200m, overgrown plantations, tall secondary growth. Occurs from India to Java and Borneo; resident.

WHITE-CROWNED SHAMA *Copsychus stricklandii* 20–27cm

Within Borneo a few endemics are found only in Sabah; some of them are on the borderline between full species and geographical races, including this shama. Differs from White-rumped Shama only in having entire centre crown pure glistening white. Intermediates occur between the two forms, but the hybrid zone is restricted and implies some reproductive barrier. Song includes loud cuckoo-like couplets; territorial, living in lower and middle storey. Found in lowland and hill forest to 1,200m, tall secondary growth. Occurs only in eastern Sabah, with intermediates in central to western Sabah.

RUFOUS-TAILED SHAMA *Trichixos pyrropyga* 22cm

This species is both less conspicuous and less common than the White-rumped Shama, but is not rare in suitable lowland habitat. Male dark grey on head, breast and wings, with a white fleck before eye; bold orange abdomen, rump and most of tail; last third of tail dark grey. Female similar but grey-brown without white fleck before eye, breast rufous-buff with faint dark band, abdomen buff. Distinguished from White-rumped Shama by shorter, rufous and grey tail, lack of white rump. Song much less varied, slow glissading whistles. Found in lowland and hill forest to 900m. Occurs in the Malay Peninsula, Sumatra and Borneo; resident.

CHESTNUT-NAPED FORKTAIL *Enicurus ruficapillus* 20cm

Single birds and pairs frequent territories along water, and can be repeatedly encountered at the same spot, day after day. Largely black and white, with white rump, deeply forked black tail with white edges and tips, and chestnut extending from crown to hind neck (in male) or entire back (in female); wings black with white cross-bar, breast white with fine black scales. Distinguished from other forktails by chestnut head colour and small size. Flies repeatedly away from approaching watcher, calling. Found on ground along streams within lowland forest up to 900m. Occurs in the Malay Peninsula, Sumatra and Borneo; resident.

EASTERN STONECHAT *Saxicola maurus* 14cm

Migrants are most conspicuous for a short time after they arrive in Borneo, and may belong to one of several different geographical races. Male has a black head, dark brown upperparts mottled with black, pale rump and white patches on the sides of neck and on wings; ochre-yellow below. Female is similar but head paler, contrasts less with the body, and underparts buff. Perches erect on tops of low bushes, dives down, gives clicking scolding calls. Found in open wasteland, scrub, disturbed vegetation along riversides in coastal plains. Occurs from eastern Europe to the Philippines; occasional migrant.

BORNEAN WHISTLING-THRUSH *Myophonus borneensis* 23cm

Dark and romantic birds, evocative of deep shady gullies and misty mornings, they give an eerie, ventriloquial whistle. Male plain very dark slaty blue, slightly iridescent purplish above. Female dark chocolate-brown, juvenile similar with faint off-white streaks on lower breast and abdomen. Black bill and legs. Hops and trots along ground in bursts with pauses, obtaining worms and insects from leaf litter, often fanning the lowered tail; call a ventriloquial sustained whistle. Found in lowlands especially around limestone and cave mouths, up into montane forest to 2,800m. Newly recognized as endemic to Borneo; resident.

CHESTNUT-CAPPED THRUSH *Zoothera interpres* 15cm

This striking small thrush is very elusive leading to it being seriously under-recorded. Mainly dark slaty grey including back, tail, wings and breast; speckled on the face and white abdomen, with two white wingbars or wing patches; the entire crown is dark chestnut. The juvenile has the crown streaked rufous and white, rufous breast with black spots, and rufous wingbars. Secretive bird, seemingly rare but it is widespread; look for it on the ground in morning and evening half-light. Found in lowland and hill forest mainly below 900m. Occurs from the Malay Peninsula to Borneo, Java and the Philippines; resident.

ISLAND THRUSH *Turdus poliocephalus* 23cm

This chunky thrush is one of the few birds that is not only found in the treeline vegetation of the highest mountains but is found more frequently there than lower down. Dusky brown to blackish head, breast, wings and tail; rusty abdomen and flanks; bright yellow bill, feet and eye-ring. Female similar to male but cut-off from dark breast to rusty abdomen is more ragged, spottier. Typically seen singly, in crowns of low trees. Found in upper montane forest and alpine zone of taller mountains, mainly above 2,100m. Occurs from Sumatra and Taiwan through west Pacific islands; in Borneo endemic race *seebohmi* is a montane resident.

WHITE-CHESTED BABBLER *Trichastoma rostratum* 13cm

This small babbler is the palest in colouring and one of the more difficult species in its genus to track down. Brown above, the rump and tail brighter rufous than back and wings; white below, the flanks washed with pale grey; bill dark above and light grey beneath, and sturdy legs pale flesh-coloured. Close to ground in dense undergrowth, singly or in pairs, or small parties, often near riverbanks in dense tangle; call up to six notes, consistently with a three-note sequence in middle, *up down down*, with single-note response by female. Found in lowland riverine forest, occasionally mangroves. Occurs in the Malay Peninsula, Sumatra and Borneo; resident.

HORSFIELD'S BABBLER *Malacocincla sepiaria* 14cm

Though it can be approached quite closely, this little bird keeps to dark undergrowth, moving ahead of the observer but giving its location away by calling. Dark brown above, the head and especially cheeks grey, and throat white. Underparts pale, the breast faintly streaked with grey, becoming fulvous on the flanks and abdomen; flesh-coloured legs. Tail slightly longer and bill heavier than in Short-tailed Babbler, which has moustache and plain breast. Call of three notes, *chek chak chooee*; often singly in undergrowth of valley bottoms. Found in lowland forest up to approximately 900m. Occurs in the Malay Peninsula, Sumatra, Borneo and Java; resident.

SHORT-TAILED BABBLER *Malacocincla malaccensis* 14cm

This is one of the more approachable forest babblers, which keeps to the ground. Dark brown above, the head and particularly cheeks grey, separated from white throat by obscure blackish moustache. Underparts pale, becoming fulvous on flanks and abdomen; flesh-coloured sturdy legs, and short squared tail; entire plumage rather smooth in appearance. Call about five notes, each of same descending pitch, or several notes each one lower pitched than the last; often seen singly, on and near ground in thick undergrowth. Found in lowland forest to about 900m. Occurs in the Malay Peninsula, Sumatra and Borneo; resident.

SOOTY-CAPPED BABBLER *Malacopteron affine* 16cm

One of the least distinctive grey tree-babblers is probably best identified by its call. Off-white below with grey tinge on breast; pale grey face, crown slightly darker grey to blackish, the upperparts greyish-brown. Distinguished from similar babblers by lack of rufous on crown, lack of dark moustache. Call a series of whistles, spaced and deliberate, going up and down the scale, rather human-like; often in mixed flocks with other birds including babblers, or in pairs in lower storey. Found in forest of lowlands and foothills. Occurs in the Malay Peninsula, Sumatra and Borneo; resident.

RUFOUS-CROWNED BABBLER *Malacopteron magnum* 18cm

A species pair is formed by the Scaly-crowned and the Rufous-crowned Babbler, the latter seldom being distinguishable on size alone. Slightly bigger, dark brown above; underparts pale, washed grey-brown on breast and flanks, with obscure streaks on breast; crown dark, sharply cut off from the rufous-chestnut forehead and forecrown. No black scales on forecrown. Call is a series of ascending, then descending, then ascending thin whistles. Found in lowland forest and secondary growth up to about 500m, in the middle and lower storeys. Occurs in the Malay Peninsula, Sumatra and Borneo; resident.

STRIPED WREN-BABBLER *Kenopia striata* 14cm

This babbler, though not often seen, appears to be common particularly in forest in the extreme level lowlands. White face with a buff patch between bill and eye, white underparts becoming progressively more buff towards under tail-coverts; crown black with white scales or streaks, but becoming brown with white streaks on the back, brown tail. Call a three-note whistle *Be my guest*, given on and close to ground in low-lying areas, swampy undergrowth. Found in extreme lowland forest, seldom in hills. Occurs in the Malay Peninsula, Sumatra and Borneo; resident.

GREY-THROATED BABBLER *Stachyris nigriceps* 13cm

One of the few birds that is confined to middle altitudes rather than being overtly montane, this babbler is extremely common at the appropriate heights. Plain fulvous brown, rufous-buff below; head pattern rather blurred including black and white streaked crown, white moustache bordered above by black and below by blackish throat; sides of face fulvous. Voice not distinctive but helpful in following flocks, a chattering and buzzing; in small parties moving through dense undergrowth, especially at sides of tracks and roads. Found in hill and montane forest about 500–1,200m. Occurs from the Himalayas to Sumatra and Borneo; resident.

CHESTNUT-WINGED BABBLER *Stachyris erythroptera* 13cm

Typical features of the *Stachyris* babblers include rufous plumage, black feathering somewhere on the head, and bright blue skin about the eye or throat. This species has plain dark grey head and breast, paling onto abdomen; bare blue skin in front of and round eye; chestnut wings and tail. Similar Grey-headed Babbler *S. poliocephala* has less extensive, streaked grey head, rufous breast. Call a quick series of *poop* notes, descending in pitch towards end. Found commonly in lowland forest and forest edge, logging tracks to about 900m, in understorey. Occurs in the Malay Peninsula, Sumatra and Borneo; resident.

BOLD-STRIPED TIT-BABBLER *Macronous bornensis* 12cm

This brown-and-buff bird is most likely to be seen in disturbed habitats rather than forest. It is small with a chestnut crown, sparsely feathered face and pale underparts heavily streaked with brown-black; back olive-brown, more rufous on the wings and tail. Two familiar calls: a repeated *chonk chonk chonk* often in clusters of three or four notes, and a harsh *sheet-sheee* with emphasis on second note. Often in small trees, feeding on insects. Found in forest edge, secondary growth, casuarina and bamboo in lowlands. Resident, recently split as Bornean endemic from Striped Tit-babbler *M. gularis* which occurs from Nepal to China and Malay Peninsula.

SUNDA LAUGHINGTHRUSH *Garrulax palliatus* 25cm

Not so conspicuous as the Chestnut-hooded Laughingthrush, this species is probably just as common but quieter. Forequarters grey, from head to upper back and lower breast, the face darker, almost black; hindquarters chestnut, including wings, tail and abdomen; a small patch of bare sky-blue skin around eye; dark bill and legs. In small flocks, giving mewing contact calls and occasionally raucous whistles, foraging in lower storey and sometimes on ground. Found in hill and montane forest from about 500 to 2,000m. Occurs only in Sumatra and Borneo; resident.

CHESTNUT-HOODED LAUGHINGTHRUSH
Rhinocichla treacheri 22cm

This laughingthrush is the most likely to be seen, noisy and sociable but elegant in appearance. Ashy grey with chestnut crown, dab of buff beneath eye and whitish streaks on forehead, and white wing panel formed by pale edges to base of primaries. The underparts are brownish-grey, under tail-coverts chestnut; bill bright orange. Call a repeated two- or three-note whistle *too-tuioo*; moving in small parties, often with other species, in middle and lower storey and edge vegetation. Found in montane forest above 900m. Endemic. Split from Chestnut-capped Laughingthrush of Malay Peninsula and Sumatra.

WHITE-BROWED SHRIKE-BABBLER *Pteruthius flaviscapis* 16cm

Fairly common in montane forest where it can be found in stunted elfin trees as well as tall forest, this species is bigger and stouter than other shrike-babblers. The male is black above including crown and sides of face, with a white eyebrow; it has a rufous-buff patch on tips of secondaries, and faint white tips to the primaries; below, it is a very smooth-looking greyish cream. Female is similar but with a washed-out appearance, olive-green wings with rufous patch on secondaries. Found in montane forest above about 800m. Occurs from Pakistan to Borneo and Java; resident.

WHITE-BELLIED ERPORNIS *Erpornis zantholeuca* 12cm

A small warbler-like bird, formerly placed in the yuhina family. Peaked crest and entire upperparts yellowish-green, face and underparts nearly white with yellow under tail-coverts. No eyebrow, wingbars, rump or tail markings and presence of crest distinguish it from warblers. Singly and in mixed foraging flocks, from canopy to lower storey; call a high-pitched, three-note *churr*. Found in forest, forest edge to about 1,000m, higher on isolated peaks. Occurs from the Himalayas to Taiwan, Sumatra and Borneo; resident.

CHESTNUT-CRESTED YUHINA *Staphida everetti* 12cm

A common species, this charming and sociable little bird is in continual motion. Small, brownish-grey on the back and wings, white throat and ring round eye, underparts and outer tail feathers; head and face chestnut, with a peaked chestnut crest on the crown. In small flocks, twittering to keep in contact, foraging amongst leaves at tips of twigs in small trees; members of party follow one another from tree to tree. Found in montane forest and forest edge, and in places down into lowland forest. Occurs only in Borneo; resident.

SUNDA BUSH-WARBLER CETTIA VULCANIA 13CM

The birds of this species on Bornean mountains differ from those in Sumatra and Java by the regular dusky markings on the underparts. Rather nondescript, dark brown with a pale buffish-white eyebrow, pale underparts, moderate-length tail; bill dark above and yellowish beneath, legs fleshy brown. Friendly Bush-warbler *Bradypterus accentor* is darker above and below, with a more heavily blotched throat, rufous eyebrow and dark greyish legs. Found in montane forest. Occurs from Sumatra and Borneo to Timor; resident.

ORIENTAL REED-WARBLER *Acrocephalus orientalis* 19cm

Though rather featureless in appearance, there is no similar bird within its habitat that could be mistaken for this species. A rather big warbler, plain brown with a buff eyebrow, whitish-buff below becoming rusty buff on the flanks; dusky colour between eye and bill forming lower border to the eyebrow. Call a series of loud harsh notes, a few croaks and a few high notes interspersed; coming briefly into view at top of reeds before it plunges back into cover. Found in reed beds, grass and scrub, especially near old mining pools, rivers, canals. Occurs through east Asia to the Philippines and Sulawesi; migrant.

YELLOW-BELLIED PRINIA *Prinia flaviventris* 14cm

All Prinias in Borneo, despite confusing variation, are of this species; few are so brightly coloured as shown. Slaty-grey head, often with a whitish brow in front of eye; olive upperparts, and cream breast grading into yellow or yellowish abdomen and under tail-coverts; narrow long tail. Distinguished from tailor birds by grey head, from Rufescent Prinia by yellowish underparts and olive (not rufous) tone to upperparts. Found in tall grass, roadside scrub, agricultural land. Occurs from Pakistan to Borneo and Java; resident.

MOUNTAIN TAILORBIRD *Phyllergates cucullatus* 12cm

This is a common bird of hill stations that must be separated from Yellow-bellied Prinia. Adult has chestnut crown, pale eyebrow, greyish nape and face grading into olive upperparts and whitish breast; entire abdomen, flanks and under tail-coverts yellow. Juvenile has olive-green crown. Distinguished from Yellow-bellied Prinia by chestnut crown of adult, olive (not grey) crown of juvenile; separated from Yellow-breasted Warbler *Seicercus montis* by lack of black stripes on crown. Found in forest, forest edge, bamboo and roadside scrub in mountains above 1,000m. Occurs from the east Himalayas to Indochina, Borneo and Java; resident.

DARK-NECKED TAILORBIRD *Orthotomus atrogularis* 11cm

Like the other tailor birds, this species has a voice out of all proportion to its size. Olive-green above, pale cream to dusky below with yellow under tail-coverts. Male has entire crown chestnut, and blackish throat and sides of neck; both sexes have yellow beneath tail and lack chestnut on thighs. Call a rolling, nasal *tttrrrrrit, tttrrrrrit*, repeated with variations; keeping to low growth along logging tracks, riverside growth. Found in forest edge and disturbed forest often with bamboo or other invasive plant growth. Occurs from north-east India to Sumatra and Borneo; resident.

ASHY TAILORBIRD *Orthotomus ruficeps* 12cm

The best chances of locating this fairly common grey and chestnut tailor bird are in mangroves. Adult has the entire head and face chestnut, nape and body grey becoming whitish on abdomen and under tail-coverts. Female paler below than male, and juvenile paler with whitish throat; juvenile distinguished from Rufous-tailed Tailorbird *O. sericeus* by lack of chestnut in tail. Call a fizz followed by a three-in-one trill, *pf-trt*, and a two-note trill *tree-dip*. Found low down in mangroves, growth behind mangroves, scrub, riverside vegetation. Occurs from southernmost Burma to Palawan and Java; resident.

ARCTIC WARBLER *Phylloscopus borealis* 12cm

Breeding widely across the northern Old World, migrants in winter concentrate in South-east Asia. Dusky olive colour above, with a pale yellowish eyebrow but no crown stripe, a single faint wingbar (a second sometimes just visible further forwards); dull creamy white below with brownish-olive flanks. No pale rump patch. Call a brief rattle, rising at the end, or a single hoarse *cheet*. Found in crowns of low trees in forest edge, secondary growth, overgrown plantations, migrating over forest from lowlands to 1,500m. Occurs across Eurasia to Alaska, south to Sulawesi in winter; migrant.

MOUNTAIN LEAF-WARBLER *Phylloscopus trivirgatus* 11cm

This is a common and brightly coloured bird of the mountains. A small warbler, olive-green back and wings without wingbars, yellow beneath from throat to under tail-coverts; bright yellow stripe down centre of crown and yellow eyebrows; blackish stripe separating crown stripe from brows, and another through eye. This bold head pattern distinguishes it from all other yellow-and-green species including Yellow-bellied Warbler and Bornean Mountain Whistler in same habitat; Kinabalu birds are duller than those from elsewhere. Found in montane forest and forest edge above 800m. Occurs from the Malay Peninsula to Java, Bali, Palawan; resident.

YELLOW-BREASTED WARBLER *Seicercus montis* 10cm

This is a rather quick-moving bird, keeping mostly to tangled growth and therefore hard to see well, although brightly coloured. Small; olive-green back, tail and wings with two obscure yellow wingbars; yellow underparts and yellow rump; crown and sides of face chestnut, with strongly marked black eyebrow. Song is a series of high, thin notes, spaced by pauses, up and down scale. Found in lower and middle storey of montane forest, scrub, ferns and often bamboo at forest edge, mainly 800–2,200m. Occurs from the Malay Peninsula to the Lesser Sunda Isles and Palawan; resident.

ASIAN BROWN FLYCATCHER *Muscicapa dauurica* 14cm

A range of small, similar-looking, greyish-brown flycatchers occurs in the region during migration. This one is greyish-brown above, unmarked whitish below with a grey wash on breast and flanks; whitish eye-ring. Bill dark with paler yellowish base to lower mandible; feet dark. The bill colour and paler breast and flanks distinguish from the Dark-sided Flycatcher *M. sibirica*. Perches upright on low trees, twigs, sallying after insects. Found in forest edge, wooded gardens and mangroves. Occurs in north-east Asia, migrating to the entire south and east of Asia; various populations migrate here.

INDIGO FLYCATCHER *Eumyias indigo* 14cm

Along tranquil, forest-lined roads, this is a common bird and can be a vocal one. It is bright indigo-blue, almost black on lores and sides of face, with forehead, lower breast and abdomen whitish; the intensity of the blue appears variable according to the light conditions. Bill and feet grey. Juveniles are washed with warm brown on breast. Perches rather upright, often low in forest and edge vegetation; song a repetitive squeak of about six notes. Found in montane forest above about 900m, forest edge, tangled growth. Occurs in Sumatra, Borneo, Java; resident.

LITTLE PIED FLYCATCHER *Ficedula westermanni* 11cm

Amongst the more commonly sighted flycatchers in the mountains is this tubby little bird. Male black on crown and sides of face, back, wings and tail; broad white brow, white wingbar and white at sides of base of tail; entirely greyish-white below. In same habitat as White-browed Shrike-babbler, yet distinguished by long eyebrow, lack of olive in wing. Female small, pale greyish-brown with rump and tail more rufous; whitish below with grey tinge to breast; faint buff wingbar. Bill short, black. Found in montane forest and forest edge above about 900m. Occurs from India and China to Sulawesi and the Lesser Sundas; resident.

BORNEAN BLUE FLYCATCHER *Cyornis superbus* 15cm

This is one of the brightest of eight blue and orange flycatchers in Borneo, and is the only species endemic to the island. Male blue above, shining blue on forehead, eyebrow and nape, and on rump; rufous-orange below, including the chin and all the way down to under tail-coverts. Female brown above, bright rufous on forehead and rump, brown wash across breast, and buffish-white chin and abdomen. Found in middle storey of hill forest about 600–1,600m, and occasionally down to lowland forest especially near rivers. Occurs only in Borneo; resident.

BLUE-AND-WHITE FLYCATCHER *Cyanoptila cyanomelana* 17cm

Male (above); female (right)

In this flycatcher, the male and female are completely different in appearance. Male bright and somewhat glossy blue above from the forehead to tip of the tail, which has two white patches at sides of base (best seen if tail spread); face and breast black, sharply cut off from the white lower breast and abdomen. Female ashy grey-brown, grading into white on the throat and the abdomen; plumage rather smooth-looking. Found in middle and upper storey of lowland and montane forest, sometimes tall secondary forest. Occurs in north-eastern Asia, migrating south to Sumatra, Borneo and Java; a non-breeding migrant, commoner in north than in Kalimantan.

GREY-HEADED CANARY-FLYCATCHER
Culicicapa ceylonensis 12cm

Common in forest but often overlooked, this flycatcher vaguely resembles a warbler because of its yellowish colouring. Head and upper breast ashy grey; back, wings and tail olive-green, lower breast and abdomen dull yellow. Rather erect posture and flycatching habit; call about six notes in three couplets, rising at the end; seen singly, in pairs and in mixed foraging flocks in middle and lower storey. Found in lowland and hill forest, tall secondary forest to about 1,000m. Occurs from Pakistan and India to Borneo and Java; resident.

GOLDEN-BELLIED GERYGONE *Gerygone sulphurea* 9cm

This tiny, light yellow bird is hard to see but its call, once learned, reveals it in a wide variety of habitats. Very small, olive-brown above, short rounded tail with pale subterminal mark on each feather; very bright yellow underparts, paler area between bill and eye. Call a series of hesitant scratchy whistles going up and down the scale, heard at all times of day; restless, quick-moving, often in canopy. Found in lowland and hill forest to about 800m, forest edge, plantations, mangroves, parks, gardens and roadside trees. Occurs from the Malay Peninsula to the Philippines and Sulawesi; resident.

RUFOUS-WINGED PHILENTOMA *Philentoma pyrhoptera* 18cm

This flycatcher-like bird has two colour phases. Male either powder-blue on head, with chestnut wings and tail, pale abdomen, or (as shown here) pale powder-blue all over, paler below; distinguished from female Maroon-breasted Philentoma by lack of dark face. Female has greyish-brown head and breast, rufous wings and tail. Both sexes have reddish eyes, rather slim body. Seen singly, in pairs or mixed foraging flocks, rather sporadic. Found in middle and lower storey of lowland and hill forest to about 900m. Occurs in the Malay Peninsula, Sumatra and Borneo; resident.

MAROON-BREASTED PHILENTOMA *Philentoma velatum* 20cm

This medium-sized but stocky bird is one of two Bornean species of philentoma. The male is overall a deep matt blue, sides of face and throat black, merging into maroon breast patch. Female similar but duller blue all over, dark sides of the face and throat; rump, abdomen and under tail-coverts are bluish-grey. Both sexes have red-brown eyes. Call a musical single whistle and a harsh churr; in middle and lower storeys. Found in lowland and hill forest up to 900m. Occurs in the Malay Peninsula, Borneo, Sumatra and Java; resident.

BLACK-NAPED MONARCH *Hypothymis azurea* 16cm

This moderate-sized flycatcher makes a small, inaccessible nest hanging from creepers. Male bright blue on head and breast, grading into duller blue on wings and tail, paler whitish abdomen; black patches on nape, round base of bill and band across breast. Female less intense blue, greyish-brown back, wings and tail, no black patches on head or breast; slender body, rather long, square-ended tail. Found in lowland and hill forest to about 1,100m, forest edge, overgrown plantations. Occurs from India to the Philippines and Java; resident.

ASIAN PARADISE-FLYCATCHER *Terpsiphone paradisi* 22–40cm

Male (above); female (right)

One of the few South-east Asian birds that has very distinct colour phases, the Asian Paradise-flycatcher is widespread in lowland forest. Male has black head and shaggy peaked crest, bright blue skin round eye, blue bill. Body and very long ribbon tail either entirely white, with black wing feathers, or bright chestnut-brown body and tail, greyish face and breast, paler abdomen. Female like the chestnut-phase male, shorter tail, shorter crest. Call a repeated harsh *chack*; in middle and lower storey, bamboo and dense undergrowth. Found in lowland and hill forest to about 1,100m, overgrown plantations, secondary forest. Occurs from Afghanistan to Borneo and Java; resident.

PIED FANTAIL *Rhipidura javanica* 18cm

The distinctive skittish behaviour of fantails has led to their Malay name of 'mad flycatcher'. Dark grey-brown above, with sharply defined short white eyebrow and white throat; a black breast-band separating the white throat from the whitish buff lower breast and abdomen; long wedge-shaped tail blackish with white tips to feathers. Seen singly or in pairs, flitting between perches while flicking and fanning tail, calling awhile with harsh churrs and snatches of song. Found in scrub, mangroves, overgrown plantations and gardens. Occurs from Peninsular Malaysia to Philippines; resident.

BORNEAN WHISTLER *Pachycephala hypoxantha* 16cm

Although quite a small bird, the head is robust in proportion to the body and the bill robust in proportion to the head, resulting in the uncomplimentary alternative name of Thickhead. Olive-green above, bright yellow sides of face and entire underparts from throat to under tail-coverts; no wingbars or tail markings; blackish lores and dark bill and legs. Female duller than male on throat and breast. Seen singly or in mixed feeding flocks, quite common in small trees along forest edge, or seeking insects in tall scrub. Found in montane forest, forest edge 900–2,500m. Occurs only in Borneo; resident.

VELVET-FRONTED NUTHATCH *Sitta frontalis* 12cm

Once a good view of this creeping bird has been obtained, it is unmistakable. Entirely purplish-blue above, whitish to pearly grey below, with bright red bill and dark feet. A black velvety patch of feathers on the forehead, extending back over eye in the male. Small, often solitary but sometimes seen in mixed feeding flocks, creeping down living and dead tree trunks and branches, in the upper and middle storeys. Found in lowland and hill forest and forest edge up to about 1,000m. Occurs from India to the Philippines and Java; resident.

YELLOW-BREASTED FLOWERPECKER
Prionochilus maculatus 9cm

This tiny bird is one of the commonest flowerpeckers in undisturbed forest. Adult mainly olive-green, with short tail and short thick bill; underparts yellow in centre from breast to abdomen, whitish with dark olive streaks at sides; small yellow and orange patch on centre of crown seldom visible. Juvenile dark olive above, yellowish below, difficult to identify in field. Singly or in pairs, call not distinctive. Found in lowland and hill forest, forest edge, tall secondary vegetation, in middle and lower storey. Occurs in the Malay Peninsula, Sumatra and Borneo; resident.

CRIMSON-BREASTED FLOWERPECKER
Prionochilus percussus 10cm

This is a flowerpecker mainly of disturbed habitats, found along logging tracks in the lowlands. The male is a greyish-blue above from crown to tail, yellow below with a red patch on the centre of breast; red patch on the top of crown and obscure white moustache streak. Female is greyish-olive, paler and greyer below with yellowish centre down breast and abdomen; obscure orange patch on crown (hard to see) and whitish moustache streak. Found in lowland forest, forest edge in understorey. Occurs in the Malay Peninsula, Sumatra, Borneo, Java.

YELLOW-RUMPED FLOWERPECKER
Prionochilus xanthopygius 9cm

Care is needed in distinguishing this Bornean endemic from the very similar Crimson-breasted Flowerpecker. Male like that species, greyish-blue above and yellow below but with small bright yellow rump patch; red patch on breast is smaller, and lacks obscure white moustache streak. Female differs from Crimson-breasted in having yellow rump but no white moustache streak. In pairs or small parties in thick tangled undergrowth. Found in lowland and hill forest, forest edge, in places up to about 1,200m. Occurs only in Borneo and neighbouring Natuna islands; resident.

ORANGE-BELLIED FLOWERPECKER
Dicaeum trigonostigma 8cm

This flowerpecker has a wide altitudinal range, being found along the edges of most forest types. Male greyish-blue on head, upper back, wings and tail; pale grey throat grading into bright orange breast and abdomen, orange lower back and rump. Female light olive with pale grey throat merging into pale yellow abdomen, yellow to orange rump. Bill slightly more slender than other flowerpeckers. Seen singly or in pairs, low down in understorey and along forest edge. Found in lowland forest, secondary growth to at least 1,000m. Occurs from east India to the Philippines; resident.

BLACK-SIDED FLOWERPECKER *Dicaeum monticolum* 8cm

Males are easy to identify but females are difficult, and best determined by being with the male. Male dark bluish above, with whitish chin and rosy scarlet patch on throat and upper breast; this patch surrounded by dark grey sides of face, lower breast and flanks, grading into buffish abdomen. Female greyish-brown all over, rather featureless, small. Seen singly or in pairs, keeping to middle and upper storey in forest but low down along forest edge; chipping call when flying from tree to tree. Found in montane forest, forest edge; 1,200m upwards. Occurs only in Borneo; endemic resident.

SCARLET-BACKED FLOWERPECKER *Dicaeum cruentatum* 8cm

One of the commoner flowerpeckers in rural habitats, this is a familiar garden bird. Male tricolored: brilliant scarlet from forehead to rump; black from sides of face down neck and wings to tail; whitish-grey from throat down centre of breast and abdomen. Female olive-grey with dark tail, bright red rump, pale whitish-grey below. Seen singly or in pairs, from low bushes to crowns of trees. Found in gardens, towns, cultivated areas, secondary vegetation from lowlands to about 1,300m. Occurs from India to south China, Sumatra and Borneo; resident.

BROWN-THROATED SUNBIRD *Anthreptes malacensis* 13cm

Male (above); female (right)

The Brown-throated Sunbird is one of the larger and commoner sunbirds to be found in gardens and open country. Male has iridescent green crown and back, purple rump and dark tail; sides of face olive. Throat dull brownish in centre with purple at sides; rest of underparts bright yellow. Female has bright greenish-yellow underparts, no white markings in square-ended tail, no distinctive rump patch. Found in gardens, plantations, secondary vegetation, open scrub, mangroves. Occurs from Burma to Indochina, the Philippines and Sulawesi; resident.

PURPLE-NAPED SUNBIRD *Hypogramma hypogrammicum* 15cm

Although not brightly coloured, this is one of the more commonly noticed sunbirds in closed-canopy forest. Deep olive-green above, darkening towards tail which has pale outer corners; below dirty yellow with heavy blackish-olive streaks from chin to abdomen; male with a rich iridescent purple patch on nape, and another on rump, not easy to see in poor light. In lower storey, sometimes accompanying mixed foraging flocks, call a single harsh note. Found in lowland and hill forest to about 900m. Occurs from south China to Sumatra and Borneo; resident.

OLIVE-BACKED SUNBIRD *Cinnyris jugularis* 12cm

This, the smallest of the sunbirds, is abundant and easily recognized in open country. The male is olive-brown above, with a glossy purplish-black throat and upper breast; the remainder of underparts bright yellow. Female small, olive above and bright yellow below including yellow sides of the face; tips of the tail feathers are dull white, best visible from below. Seen singly or in pairs, foraging on bushes and low trees, often briefly on exposed perches. Found in gardens, secondary vegetation, any open country and mangroves. Occurs from south China to the Philippines and through to Australia; resident.

CRIMSON SUNBIRD *Aethopyga siparaja* 10–12cm

Of several bright red sunbirds in South-east Asia, this is the most likely to be seen in disturbed country. Male predominantly bright red with crimson back, yellow rump, dark green tail (often appearing black) with elongated central feathers, and iridescent green forehead; abdomen grey, wings dark. Female entirely dull olive-green, paler yellow below, dull greenish under tail-coverts; distinguished from the female Brown-throated by duller underparts and small size. Found in forest edge, tall secondary growth and plantations. Occurs from India to the Philippines and Sulawesi; resident.

TEMMINCK'S SUNBIRD *Aethopyga temminckii* 10–12cm

Very similar to the Crimson Sunbird in appearance, this species is its equivalent in hill and montane forest. Male is predominantly bright red, scarlet tail separated from back by blackish upper tailcoverts, narrow yellow rump. The iridescent violet crown and eyebrow often appear to be black. Female is olive-green like Crimson Sunbird but head greyer, wings and tail tinged reddish, flanks tinged grey. Found in hill and lower montane forest to about 1,500m, forest edge and along roadsides in montane areas. Occurs in the Malay Peninsula, Sumatra and Borneo, not in Java.

LITTLE SPIDERHUNTER *Arachnothera longirostra* 16cm

The Little Spiderhunter is the commonest of its group and typical of disturbed forest, where it feeds at banana inflorescences. Olive-green above, greyish on head with obscure moustache streak, throat whitish grading into olive breast, and yellow abdomen and under tail-coverts. Little orange tufts occasionally visible at bend of wing, more often present in males than females. Call a loud *tchek*, at intervals during rapid direct flight and sometimes monotonously when perched. Found in forest, logged and secondary forest to about 1,500m. Occurs from India to Java and the Philippines; resident.

LONG-BILLED SPIDERHUNTER *Arachnothera robusta* 21cm

The plumage is not very distinctive, but this species has the most impressive bill of all the spiderhunters. Large, with long, thick black bill; dark olive-green patternless head, olive-green above, yellowish and faintly streaked below, becoming yellow on abdomen. Tail feathers with pale tips on underside. Usually feeds high in forest canopy, often perched on bare twigs; protects food sources aggressively; call a single harsh *chack*. Found in lowland, hill and lower montane forest up to 1,300m, commoner towards the upper end of this range. Occurs in the Malay Peninsula, Sumatra, Borneo and Java; resident.

SPECTACLED SPIDERHUNTER *Arachnothera flavigaster* 21cm

Bulkier than the Yellow-eared Spiderhunter, this species often feeds lower down in the trees. Broad, complete yellow ring round eye, often linking up with small yellow ear patch; plumage olive above, pale olive-grey below without streaks; large size. Takes nectar from flowers of a variety of forest trees, often protecting food sources aggressively. Found in logged forest and secondary growth, coconut plantations and abandoned cultivation, mainly lowlands but occasionally over 1,000m. Occurs in the Malay Peninsula, Sumatra and Borneo; resident.

YELLOW-EARED SPIDERHUNTER
Arachnothera chrysogenys 17cm

Since this bird is difficult to distinguish from the Spectacled Spiderhunter, special attention must be paid to its head pattern. Often incomplete yellow ring round eye, typically not linking up with yellow ear patch; plumage olive above, pale olive-grey below appearing faintly streaked, becoming yellowish on abdomen and beneath tail. Juvenile has eye-ring but ear patch is reduced. Feeds singly, mostly high up in forest canopy. Call when it is perched is a slow trill ending in a single long note. Found in lowland forest, forest edge, secondary vegetation, gardens up to 900m. Occurs in the Malay Peninsula, Sumatra, Borneo, Java; resident.

STREAKY-BREASTED SPIDERHUNTER
Arachnothera affinis 17cm

A soberly plumaged bird, dull olive green above, very hard to separate from Grey-breasted Spiderhunter *A. modesta*. This species is slightly bigger, with cool grey underparts heavily streaked all the way up to the throat and chin. Identified by cold colouring, streaking, and lack of other features; cool yellowish patch between legs might be seen. Juvenile without streaks on breast. Found in forest, in lowlands and highlands of Sabah, in highlands elsewhere on the island. Occurs only in Borneo, Java and Bali; resident.

ORIENTAL WHITE-EYE *Zosterops palpebrosus* 11cm

This is a popular cage-bird which is widespread in forest, overlapping with similar Everett's White-eye *Z. everetti* in the mountains. Bright olive-green above, bright yellow on throat merging into grey abdomen and under tail-coverts; wings and tail darker. Bright white feathering round eye, set off by darker sides of face; forehead often with traces of yellow. Everett's White-eye has darker upperparts, no yellow on forehead, darker grey abdomen. Found in mangroves, forest edge, logging tracks from coast up to about 1,800m. Occurs from Afghanistan to Borneo and Java; resident.

BLACK-CAPPED WHITE-EYE *Zosterops atricapilla* 10cm

This is one of the two white-eyes, quite easily distinguishable, that occur in the mountains of Borneo. Similar to lowland Oriental White-eye and montane Everett's White-eye *Z. everetti* but dark olive-green above, especially on the forehead, crown and cheeks which are dusky, blackish; olive breast grading into greyish abdomen; distinct white ring round eye. Everett's has a yellow streak down centre of abdomen, missing in this species. Small, active flocks keeping in contact by twittering, working through crowns of smaller trees for insects. Found in lower and upper montane forest, forest edge, about 900–2,100m. Occurs in Sumatra and Borneo; resident.

MOUNTAIN BLACKEYE *Chlorocharis emiliae* 12cm

Accommodating and curious birds, Mountain Blackeyes investigate people closely, and often carry on feeding and calling undisturbed by observers, even at close range. Small, active, dark olive to sage-green birds, with blackish feathering from base of bill surrounding eye, and rather long orange bill; the black eye-ring is emphasized by paler, yellowish-green feathering on sides of head. Pretty song and gives incessant fluid twittering calls *tuiu*, while foraging in understorey and stunted trees for insects, occasionally small fruits. Found in understorey of montane forest, becoming increasingly common in stunted upper montane and heath forest, 1,200–3,600m. Occurs only in Borneo; resident.

DUSKY MUNIA *Lonchura fuscans* 10cm

This is the commonest of Borneo endemics, and the first that any birdwatcher will see. Dark chocolate-brown all over, with thick bill, the lower mandible pale grey and upper mandible darker. White-bellied Munia *L. leucogastra* is similar but has white abdomen and faint whitish streaks on back. Lives in small flocks, which creep and fly in bumble-bee fashion amongst tall grass, along road margins, faintly chirping to each other; nests amongst tall grass, against embankments or in isolated small trees in open country, and often roosts in flocks inside disused nests. Found in open grassland, scrub and rice fields. Occurs only in Borneo; resident.

CHESTNUT MUNIA *Lonchura atricapilla* 11cm

The black head of this species is the best feature to look for when the birds are perched or during the whirring bee-like flight of small flocks in grassland. Black head and upper breast, sharply defined from rich brown body, wings and tail. Juvenile overall warm fawn-brown, tends to be darker on head. Bill thick, conical, bluish-grey. Forages on grass seed-heads, rice, forming big flocks. Found in gardens, scrub, secondary growth and rice fields, plantation edges and swamps. Occurs from India to the Philippines and Sulawesi, introduced elsewhere; resident.

EURASIAN TREE-SPARROW *Passer montanus* 14cm

Eurasian Tree-sparrows are excellent colonists and make their way even to remote clearings, especially in Borneo, by following rivers and riverside cultivation. Crown chestnut; throat, sides of face and ear patch black, contrasting with cold buff underparts; back and wings brown mottled with black; wings with two pale bars. Commonly feeding on ground, active even at night in bright cities. Found in all open habitats from scrub and coasts to gardens, cultivation, towns, reaching clearings and villages in mountains. Occurs throughout Europe and Asia to Sumatra and Java, introduced to Borneo, North America and Australia; resident.

ASIAN GLOSSY STARLING *Aplonis panayensis* 20cm

In towns and villages these birds can often be found on high-tension wires, but they also frequent coconut palms and bare trees. Adult entirely black with an oily green gloss all over; eyes bright red. Juvenile dark greenish-grey above, somewhat glossy, and dirty buff with grey streaks below. Slim, small-headed; flight direct, rapid wingbeats showing small triangular wings and short tail; in small flocks, often giving noisy piping notes. Found in any secondary vegetation, gardens, towns, open country including coastal areas and small islands. Occurs from India to the Philippines and Sulawesi; resident.

CRESTED MYNA *Acridotheres cristatellus* 28cm

Out of the confusing group of dark grey Asian mynas, only this one has so far been introduced to Borneo. Dark grey (nearly black) starling-like bird, short ragged crest hanging forward over red-based horn-coloured bill; white patch at base of primaries prominent in flight, narrow white tip of tail, and under tail-coverts grey with narrow white bars. No bare skin round orange eye. Sociable and rather noisy, feeding on ground, roosting on buildings and trees. Found in several larger towns, beginning to spread to suburban areas. Occurs in mainland South-east Asia; introduced resident in Kuching, Kota Kinabalu.

COMMON HILL-MYNA *Gracula religiosa* 30cm

Still fairly common in lowland forest and wooded rural areas, Hill Mynas' loud calls are unmistakable. Heavy, plumage entirely black with purplish gloss, white patches at base of primaries; bill orange to yellow, feet yellow, and flaps of bare yellow skin on face and nape. Feeds mainly on fruit in canopy of trees, call typically a ringing *tiong*, the wings thrumming in flight. Found in tall lowland forest, forest edge, isolated trees in rural areas, often perches on dead branches and nests in tree holes. Occurs from India to Borneo, Java and Palawan, introduced in some other countries; resident.

BLACK-NAPED ORIOLE *Oriolus chinensis* 26cm

One of Borneo's mystery birds, very rarely recorded in easternmost parts of Sarawak, Brunei, Kinabalu foothills and southernmost Kalimantan; all could be vagrants. Male brilliant yellow with black mask through eyes meeting at nape, black wings, black tail with yellow tips; bill pinkish-orange. Females duller, the back and underparts yellowish-green to dull olive. Juveniles duller still, olive above and off-white below with blackish streaks. Call a melodious four-note whistle, *What the devil!* with much individual variation. Found in mangrove edge, scrub, secondary growth and gardens, into city centres. Occurs from India to the Philippines; resident mainly in coastal plains, supplemented by migrants from islands to far inland.

BLACK-AND-CRIMSON ORIOLE *Oriolus cruentus* 22cm

Often seen in silhouette or in poor misty conditions, this apparently all-black bird is worth a second look. Males and old females are intense black with rounded crimson patch on the breast, and a crimson flash on primary coverts. Juveniles and young females black with the breast and abdomen greyish. Both the sexes have blue feet and pale blue bill. Call a simple wheeze or a plaintive mew, unlike bell-like notes of most other orioles. Found in tall montane forest above approximately 700m, keeping to dark middle storey. Occurs in Malay Peninsula, Sumatra, Borneo and Java; resident.

ASHY DRONGO *Dicrurus leucophaeus* 28cm

Though this drongo is a bird of open country, it might also be seen in forest-edge habitats. Plumage is usually ashy grey to dark grey, can appear almost black but always has a reddish eye; shows variety of colours owing to presence of various races on migration, some particularly pale with whitish face. Tail deeply forked, the outer tips not turned upwards. Can be separated from grey-coloured cuckoo-shrikes by forked tail and complete lack of barring. Found in mangroves, coastal scrub, secondary vegetation, forest edge, especially on west coast plain. Occurs from the Middle East to China and Java; migrant.

CROW-BILLED DRONGO *Dicrurus annectans* 27cm

One of the less common drongos, this is most likely to be seen in hill forest. Heavy, stout bill made heavier in appearance by bristles at base; tail long and forked, with the outer tips strongly turned upwards. Juvenile has whitish scaling on flanks and under wings. Calls include a series of twanging notes descending in pitch, also varied rasps and bell-like notes; usually seen singly, on passage. Found in hill forest up to 1,000m during passage, and lower down into mangroves, secondary growth, plantations during wintering period. Occurs from the Himalayas to Borneo and Java; passage migrant and non-breeding visitor.

HAIR-CRESTED DRONGO *Dicrurus hottentottus* 25cm

Also known as the Hair-crested Drongo, from the thin hairlike plumes lying over the crown of some birds, as in those from Borneo, this is one of the glossiest drongos. Glossy black all over, with iridescent spangles especially on crown and breast, the long tail forked and the blunt outer tips curved back and up; heavy bill. Seen on exposed perches, tree tops, wires, hawking after flying insects and returning to same perch; sometimes in mixed foraging flocks. Found in sparse forest and forest edge, middle storey, in lowlands and hills. Occurs discontinuously from India to Java and Bali; resident.

GREATER RACKET-TAILED Drongo *Dicrurus paradiseus* 32–57cm

The most conspicuous and well-known drongo is also the biggest in the region and the most spectacular. All black plumage, large size and somewhat rounded crown; tail forked not square, the two outer feathers having elongated bare shafts (up to 25cm) and large rounded and twisted racquets, which may be broken off. Traces of elongated broken shafts are usually visible, and large size, profile, and forked tail without any upturning at tips are then distinctive. Great variety of harsh and bubbling, bell-like notes, calls frequently. Found in lowland forest, plantations, secondary growth. Occurs from India to Borneo and Java; resident.

CRESTED JAY *Platylophus galericulatus* 32cm

Though it can move rapidly and inconspicuously through the forest, the Crested Jay's calls, once learned, reveal it to be reasonably common. Adult all dark, chocolate brown (Javan birds, shown, are blacker), darker head, with small white patch high up on side of neck, and a very long spatulate crest standing straight up over crown. Juvenile browner above, rufous tips to wing coverts, centre of belly pale to barred, and crest shorter. Call a long harsh chatter like a shrike. Found in lowland forest to about 800m, mainly middle storey. Occurs in the Malay Peninsula, Sumatra, Borneo and Java; resident.

BORNEAN BLACK MAGPIE *Platysmurus aterrimus* 40cm

Sometimes they are not seen for days on end, but at other times it seems impossible to escape from these birds in the forest. A black crow-like bird with heavy bill and short bunchy crest; Bornean Black Magpie is distinctive in lacking the white wing patch characteristic of the Black Magpie *P. leucopterus* of mainland South-east Asia. Juveniles lack the crest. Call a sheep-like bleat, repeatedly; also a truly remarkable range of bell-like notes, bak-bong, with many variations. Found in middle storey of lowland forest, sometimes in remnant riverine forest and overgrown plantations. Endemic, recently split from Black Magpie; resident.

COMMON GREEN MAGPIE *Cissa chinensis* 38cm

A bird of the middle slopes and lower montane zone, renowned for its emerald colouring in which the tone of yellow is rather variable. Slightly longer but slimmer than Short-tailed GreenMagpie, bright green with red bill and feet, black mask through eye; chestnut wings with black and white tips to secondaries as well as to tail feathers. Usually occurs low down in forest, in pairs or small parties, quiet for long periods but occasionally noisy harsh chatters, hisses and three-note whistles. Found in montane forest, about 900–1,800m. Occurs from the Himalayas to Sumatra and Borneo; resident.

SHORT-TAILED GREEN MAGPIE *Cissa thalassina* 35cm

The more intense green plumage, an unusual shade amongst birds, separates this species from all but the closely related Common Green. Slightly bulkier, slightly shorter-tailed than Common Green, bright green with red bill and feet, black mask through eye; chestnut wings in which tips of secondaries may be pale but lack black and white tips; graduated tail with black and white tips. Usually low down in the forest, occasionally to ground level, in small parties; sometimes giving harsh loud cries but its whistle a pale imitation of Common Green. Found in montane forest at about 900–2,200m, commoner higher up. Occurs discontinuously from China to Borneo and Java; resident.

BORNEAN TREEPIE *Dendrocitta cinerascens* 45–50cm

No other bird in montane forest is so noisy or gives such a wide range of bell-like clangs, grunts and groans as this. Rufous face and underparts, pale grey cap and back. Wings black with white patch at base of flight feathers; tail grey with black tips, very long, and flaunted when active. Long tail and calls make it unmistakable. Fluid whistles, grunts, churrs and emphatic bell-like *ka-tonk*. Feeding singly and in small parties on fruits and insects, active and noisy. Found from about 800 to 2,700m, commoner within lower parts of this range, through the mountains of Sarawak, Sabah and Kalimantan. Occurs only in Borneo; endemic montane resident.

FURTHER READING

The following books and other publications should be of interest to those wishing to learn more about birds, birdwatching and other wildlife in the region. *A Field Guide to the Birds of Borneo* by Susan Myers is the key field guide to the island.

Bransbury, J. *A Birdwatcher's Guide to Malaysia*. Waymark Publishing, Australia, 1993

Das, I. *A Field Guide to the Reptiles of South-East Asia*. New Holland Publishers, London, 2010

Das, I. *A Photographic Guide to the Snakes and Other Reptiles of Borneo*. New Holland Publishers, London, 2006

Fisher, T and Hicks, N. *A Photographic Guide to the Birds of the Philippines*. New Holland Publishers, London, 2000

Francis, C.M. *A Field Guide to the Mammals of South-East Asia*. New Holland Publishers, London, 2008

Francis, C.M. *A Photographic Guide to the Mammals of South-East Asia*. New Holland Publishers, London, 2001

Garbutt, N. and Prudente, C. *Wild Borneo*. New Holland Publishers, London, 2007

Howard, R. and Moore, A. *A Complete Checklist of the Birds of the World*. Second edition. Oxford University Press, Oxford, 1991

King, B., Woodcock, M. and Dickinson, E. *A Field Guide to the Birds of South-East Asia*. Collins, London, 1975

Myers, S. *A Field Guide to the Birds of Borneo*. New Holland Publishers, London, 2009

Robson, C. *A Field Guide to the Birds of South-East Asia*. Second Edition. New Holland Publishers, London, 2008

Smythies, B.E. 'An annotated checklist of the birds of Borneo'. *Sarawak Museum Journal*, VII (9), 1957: 523–818.

Smythies, B.E. *The Birds of Borneo*. First edition. Oliver & Boyd, Edinburgh and London, 1960

Tilford, T and Compost, A. *A Photographic Guide to the Birds of Java, Sumatra and Bali*. New Holland Publishers, London, 2000

White, T.E. *A Field Guide to the Bird Songs of South-east Asia*. National Sound Archives, London, 1984

INDEX

Aceros corrugatus 70
  undulatus 71
Acridotheres cristatellus 133
Acrocephalus orientalis 111
Actitis hypoleucos 40
Adjutant, Lesser 22
Aegithina tiphia 94
  viridissima 95
Aethopyga siparaja 126
  temmiincki 127
Alcedo atthis 65
  meninting 65
Alophoixus bres 91
  ochraceus 91
  phaeocephalus 92
Amaurornis phoenicurus 32
Anas acuta 23
Anhinga melanogaster 14
Anorrhinus galeritus 70
Anthracoceros
  albirostris 72
  malayanus 71
Anthreptes malacensis 125
Anthus richardi 83
  rufulus 83
Aplonis panayensis 133
Apus affinis 63
  nipalensis 63
  pacificus 63
Arachnothera
  affinis 129
  chrysogenys 129
  flavigaster 128
  longirostra 127
  modesta 129
  robusta 128
Ardea alba 16
  cinerea 15
  purpurea 16
Ardeola bacchus 18
  speciosa 19
Arenaria interpres 41
Argus, Great 30
Argusianus argus 30

Babbler,
  Chestnut-winged 107
  Grey-headed 107
  Grey-throated 106
  Horsfield's 104
  Rufous-crowned 105
  Short-tailed 104
  Sooty-capped 105
  White-chested 103
Barbet, Gold-whiskered 73
  Golden-naped 73
Batrachostomus, affinis 61
  auritus 61
  cornutus 61
  stellatus 61
Bee-eater, Blue-tailed 68
  Blue-throated 67
Berenicornis comatus 69
Bittern, Cinnamon 21
  Yellow 21
Blackeye, Mountain 131
Blythipicus rubiginosus 77
Boobook, Brown 60
Bradypterus accentor 110
Broadbill, Banded 78
  Black-and-red 78
  Black-and-yellow 78
  Green 79
  Long-tailed 79
  Whitehead's 80
Bubo sumatranus 59
Bubulcus coromandus 18
Buceros rhinoceros 72
Bulbul, Ashy 94
  Black-and-white 86
  Black-crested 87
  Black-headed 87
  Bornean 87
  Buff-vented 92
  Cinereous 94
  Flavescent 89
  Grey-cheeked 91
  Hairy-backed 93
  Ochraceous 91
  Olive-winged 90
  Pale-faced 89
  Puff-backed 88
  Red-eyed 90
  Red-whiskered 87
  Scaly-breasted 88
  Spectacled 91
  Straw-headed 86
  Streaked 93
  Yellow-bellied 92
  Yellow-vented 89
Bush-warbler, Friendly 110
  Sunda 110
Butorides striata 19

Cacomantis sonneratii 55
Calidris ferruginea 43
  ruficollis 42
  subminuta 42
  temminckii 42
Caloenas nicobarica 49
Calyptomena viridis 79
  whiteheadi 80
Canary-flycatcher, Grey-headed 17
Caprimulgus affinis 62
  macrurus 62
Cecropsis striolata 82
Centropus bengalensis 58
  sinensis 57
Cettia vulcania 110
Chalcophaps indica 48
Charadrius dubius 36
  leschenaultii 37
  mongolus 37
Chlidonias hybrida 44
  leucopterus 44
Chlorocharis emiliae 131
Chloropsis
  cochinchinensis 96
  cyanopogon 96
  sonnerati 95
Chrysocolaptes lucidus 76
Chrysophlegma
  mentalis 76
  mineaceus 75
Ciconia stormi 22
Cinnyris jugularis 126
Circus spilonotus 26
Cissa chinensis 138
  thalassina 139
Clamator coromandus 54
Collocalia esculenta 62
Coot, Common 34
Copsychus
  malabaricus 99
  saularis 99
  stricklandii 100
Cormorant, Great 14
Coucal, Greater 57
  Lesser 58
Crake, Band-bellied 31
  Red-legged 31
  Ruddy-breasted 31
Cuckoo, Banded Bay 55
  Chestnut-winged 54
Cuckoo-dove, Little 47
Culicicapa ceylonensis 117
Curlew, Eurasian 38
Cyanoptila cyanomelana 117
Cymbirhynchus
  macrorhynchos 78
Cyornis superbus 116

141

Darter, Oriental 14
Dendrocitta cinerascens 139
Dendrocopos moluccensis 74
Dendrocygna javanica 23
Dicaeum
    cruentatum 124
    monticolum 124
    trigonostigma 123
Dicrurus annectans 136
    hottentottus 136
    leucophaeus 135
    paradiseus 137
Dinopium javanense 76
Dollarbird 68
Dove, Emerald 48
    Peaceful 48
    Spotted 47
    Zebra 48
Dowitcher, Asian 42
Drongo, Ashy 135
    Crow-billed 136
    Greater Racket-tailed 137
    Hair-crested 136
Dryocopus javensis 77
Ducula aenea 51
    badia 52
    bicolor 52

Eagle, Black 27
Eagle-owl, Barred 59
Egret, Eastern Cattle 18
    Great 16
    Intermediate 17
    Little 17
Egretta garzetta 17
    intermedia 17
Elanus caeruleus 24
Enicurus ruficapillus 101
Erpornis, White-bellied 163
Erpornis zantholeuca 163
Eudynamys scolopacea 55
Eumyias indigo 115
Eurylaimus javanicus 78
    ochromalus 78
Eurystomus orientalis 68

Fairy-bluebird, Asian 97
Fantail, Pied 120
Ficedula westermanni 116
Fireback, Crested 29
    Crestless 29
Fish-owl, Buffy 59
Flameback, Common 76
    Greater 76

Flowerpecker,
    Black-sided 124
    Crimson-breasted 122
    Orange-bellied 123
    Scarlet-backed 124
    Yellow-breasted 122
    Yellow-rumped 123
Flycatcher-shrike,
    Bar-winged 85
    Black-winged 85
Flycatcher,
    Asian Brown 115
    Blue-and-white 117
    Bornean Blue 116
    Dark-sided 115
    Indigo 115
    Little Pied 116
Forktail, Chestnut-naped 101
Fregata andrewsi 15
Frigatebird, Christmas Island 15
Frogmouth, Blyth's 61
    Gould's 61
    Large 61
    Sunda 61
Fruit-dove, Jambu 51
Fulica atra 34

Gallicrex cinerea 32
Gallinago gallinago 41
    megala 41
    stenura 41
Gallinula chloropus 33
Gallirallus striatus 31
Garrulax palliatus 108
Geopelia striata 48
Gerygone, Golden-bellied 118
Gerygone sulphurea 118
Godwit, Bar-tailed 37
    Black-tailed 37
Gracula religiosa 134
Green-pigeon, Little 49
    Pink-necked 50
    Thick-billed 50
Greenshank, Common 39
Gull, Black-headed 43

Halcyon pileata 66
Haliaeetus leucogaster 25
Haliastur indus 25
Hanging-parrot, Blue-crowned 53
Harpactes diardii 63
    duvaucelii 64
    kasumba 64
    whiteheadi 64

Hawk-cuckoo, Large 54
Hawk-eagle, Blyth's 27
    Changeable 27
Hemipus
    hirundinaceus 85
    picatus 85
Hemixos cinereus 94
    flavala 94
Heron, Grey 15
    Little 19
    Purple 16
Hierococcyx sparverioides 54
Hill-myna, Common 134
Himantopus himantopus 35
Hirundo rustica 82
    tahitica 82
Honey-buzzard, Oriental 27
Hoopoe, Common 69
Hornbill, Black 71
    Bushy-crested 70
    Oriental Pied 72
    Rhinoceros 72
    White-crowned 69
    Wreathed 71
    Wrinkled 70
Hydrophasianus chirurgus 34
Hypogramma hypogrammicum 125
Hypothymis azurea 119

Ictinaetus malayensis 27
Imperial-pigeon, Green 51
    Mountain 52
    Pied 52
Iole olivacea 92
Iora, Common 94
    Green 95
Irena puella 97
Ixobrychus
    cinnamomeus 21
    sinensis 21
Ixos malaccensis 93

Jacana, Pheasant-tailed 34
Jay, Crested 137

Kenopia striata 106
Ketupa ketupu 59
Kingfisher,
    Black-capped 66
    Blue-eared 67
    Collared 67
    Common 65
    Stork-billed 66

Kite,
    Black-shouldered 24
    Brahminy 25
Koel, Asian 55

Lalage nigra 84
Lanius cristatus 98
    schach 98
    tigrinus 97
Larus ridibundus 43
Laughingthrush,
    Chestnut-capped 108
    Chestnut-hooded 108
    Sunda 108
Leafbird,
    Blue-winged 96
    Greater Green 95
    Lesser Green 96
Leaf-warbler, Mountain 114
Leptoptilos javanicus 22
Limnodromus semipalmatus 42
Limosa lapponica 37
    limosa 37
Lonchura atricapilla 132
    fuscans 131
    leucogastra 131
Lophura bulweri 30
    erythrophthalma 29
    ignita 29
Loriculus galgulus 53

Macronous bornensis 107
    gularis 107
Macropygia ruficeps 47
Magpie, Bornean Black 138
    Common Green 138
    Short-tailed Green 139
Magpie-robin, Oriental 99
Malacocincla malaccense 104
    sepiaria 104
Malacopteron affine 105
    magnum 105
Malkoha, Black-bellied 56
    Chestnut-breasted 57
    Chestnut-bellied 56
    Raffles's 56
Marsh-harrier, Eastern 26
Megalaima chrysopogon 73
    pulcherrima 73
Melanoperdix nigra 28
Merops philippinus 68
    viridis 67

Minivet, Ashy 84
    Grey-chinned 85
    Scarlet 85
Monarch, Black-naped 119
Moorhen, Common 33
Motacilla tschutschensis 83
Mulleripicus pulverulentus 77
Munia, Chestnut 132
    Dusky 131
    White-bellied 131
Muscicapa sibirica 115
    dauurica 115
Myophonus borneensis 102
Myna, Crested 133

Night-heron,
    Black-crowned 20
    Rufous 20
Nightjar, Large-tailed 62
    Savanna 62
Ninox scutulata 60
Nisaetus alboniger 27
    limnaeetus 27
Nuthatch, Velvet-fronted 121
Nycticorax caledonicus 20
    nycticorax 20

Onychoprion anaethetus 45
    fuscatus 45
Oriole,
    Black-and-crimson 135
    Black-naped 134
Oriolus chinensis 134
    cruentus 135
Orthotomus atrogularis 112
    ruficeps 113
    sericeus 113
Osprey 24
Otus rufescens 58

Pachycephala hypoxantha 121
Painted-snipe, Greater 35
Pandion haliaetus 24
Paradise-flycatcher, Asian 120
Parakeet, Long-tailed 53
Partridge, Black 28
    Crested 28
Passer montanus 132
Pelargopsis capensis 66
Pericrocotus divaricatus 84

    solaris 85
    speciogus 85
Pernis ptilorhynchus 27
Phalacrocorax carbo 43
Pheasant, Bulwer's 30
Philentoma,
    Maroon-breasted 119
    Rufous-winged 118
Philentoma pyrhoptera 118
    velatum 119
Phyllergates cucullatus 113
Phylloscopus borealis 113
    trivirgatus 114
Piculet, Speckled 74
Picumnus innominatus 74
Picus puniceus 75
Pigeon, Nicobar 49
Pintail, Northern 23
Pipit, Paddyfield 83
    Richard's 83
Pitta, Banded 80
    Blue-headed 81
    Blue-winged 81
    Mangrove 81
Pitta baudii 81
    guajana 80
    megarhyncha 81
    moluccensis 81
Platylophus galericulatus 137
Platysmurus aterrimus 138
Plover, Grey 36
    Little Ringed 36
    Pacific Golden 36
Pluvialis fulva 36
    squatarola 36
Pond-heron, Chinese 18
    Javan 19
Porphyrio indicus 33
    porphyrio 33
Porzana fusca 31
    paykullii 31
Prinia flaviventris 111
    rufescens 111
Prinia, Rufescent 111
    Yellow-bellied 111
Prionochilus maculatus 122
    percussus 122
    xanthopygius 123
Psarisomus dalhousiae 79
Psittacula longicauda 53
Pteruthius flaviscapis 109
Ptilinopus jambu 51
Pycnonotus atriceps 87
    brunneus 90
    erythrophthalmos 91
    eutilotus 88

143

flavescens 89
goiavier 89
leucops 89
melanicterus 87
melanoleucos 86
montis 87
plumosus 90
squamatus 88
zeylanicus 86

Rail, Slaty-breasted 31
Rallina fasciata 31
Redshank, Common 38
  Spotted 38
Reed-warbler, Oriental 111
Rhinocichla treacheri 56
Rhinortha chlorophaeus 56
Rhipidura javanica 120
Rhopodytes diardi 56
  sumatranus 56
Rollulus rouloul 28
Rostratula benghalensis 35

Sandpiper, Common 40
  Curlew 43
  Marsh 39
  Terek 40
  Wood 39
Sand-plover, Greater 37
  Lesser 37
Saxicola maurus 101
Scops-owl, Reddish 58
Sea-eagle, White-bellied 25
Seicercus montis 114
Serpent-eagle, Crested 26
  Mountain 26
Shama, Rufous-tailed 100
  White-crowned 100
  White-rumped 99
Shrike, Brown 98
  Long-tailed 98
  Tiger 97
Shrike-babbler, White-browed 109
Sitta frontalis 121
Snipe, Common 41
  Pintail 41
  Swinhoe's 41
Spiderhunter,
  Grey-breasted 129
  Little 127
  Long-billed 128
  Spectacled 128
  Streaky-breasted 129
  Yellow-eared 129
Spilornis cheela 26
  kinabaluensis 26

Stachyris erythroptera 107
  nigriceps 106
  poliocephala 107
Staphida everetti 110
Starling, Asian Glossy 133
Sternula albifrons 46
Sterna sumatrana 45
Stilt, Black-winged 35
Stint, Long-toed 42
  Red-necked 42
  Temminck's 42
Stonechat, Eastern 101
Stork, Storm's 22
Streptopelia chinensis 47
Strix leptogrammica 60
Sunbird,
  Brown-throated 125
  Crimson 126
  Olive-backed 126
  Purple-naped 125
  Temminck's 127
Swallow, Barn 82
  House 82
  Striated 82
Swamp-hen, Black-backed 33
Swift, Fork-tailed 63
  House 63
Swiftlet, Glossy 62

Tailorbird, Ashy 113
  Dark-necked 112
  Mountain 112
  Rufous-tailed 113
Tern, Black-naped 45
  Bridled 45
  Chinese Crested 46
  Great Crested 46
  Lesser Crested 46
  Little 46
  Sooty 45
  Whiskered 44
  White-winged 44
Terpsiphone paradisi 120
Thalasseus
  bengalensis 46
  bergii 46
  bernsteini 46
Thrush,
  Chestnut-capped 102
  Island 103
Tit-babbler,
  Bold-striped 107
  Striped 107
Todirhamphus chloris 67
Treepie, Bornean 139
Tree-sparrow, Eurasian 132
Treron curvirostra 50

  olax 49
  vernans 50
Trichastoma rostratum 103
Trichixos pyrropyga 100
Tricholestes criniger 93
Triller, Pied 84
Tringa erythropus 38
  glareola 39
  nebularia 39
  stagnatilis 39
  totanus 38
Trogon, Diard's 63
  Red-naped 64
  Scarlet-rumped 64
  Whitehead's 64
Turdus poliocephalus 103
Turnstone, Ruddy 41

Upupa epops 69

Wagtail, Eastern Yellow 83
Warbler, Arctic 113
  Yellow-breasted 114
Watercock 32
Waterhen, White-breasted 32
Whimbrel 38
Whistler, Bornean 121
Whistling-duck, Lesser 23
Whistling-thrush, Bornean 102
White-eye,
  Black-capped 130
  Everett's 130
  Oriental 130
Wood-owl, Brown 60
Woodpecker, Banded 75
  Checker-throated 76
  Crimson-winged 75
  Great Slaty 77
  Maroon 77
  Sunda Pygmy 74
  White-bellied 77
Wren-babbler, Striped 106

Xenus cinereus 40

Yuhina, Chestnut-crested 110

Zanclostomus curvirostris 57
Zoothera interpres 102
Zosterops
  atricapilla 130
  palpebrosus 130
  everetti 130